PUMPS

(A GUIDE FOR PUMP MANUFACTURERS)

PUMPS

(A GUIDE FOR PUMP MANUFACTURERS)

K M Srinivasan
Ex-Professor
Mechanical Engineering
PSG College of Technology
Coimbatore, India

NEW AGE
REFERENCE

An Imprint of

NEW AGE INTERNATIONAL (P) LIMITED, PUBLISHERS

LONDON • NEW DELHI • NAIROBI

Bangalore • Chennai • Cochin • Guwahati • Hyderabad • Kolkata • Lucknow • Mumbai

Visit us at **www.newagepublishers.com**

Published by New Age International (P) Ltd., Publishers
First Edition: 2017

GLOBAL OFFICES

- **New Delhi** **NEW AGE INTERNATIONAL (P) LIMITED, PUBLISHERS**
  7/30 A, Daryaganj, New Delhi-110002, (INDIA)
  Tel.: (011) 23253771, 23253472, **Telefax:** 23267437, 43551305
  Visit us at www.newagepublishers.com • e-mail: contactus@newagepublishers.com
- **London** **NEW AGE INTERNATIONAL (UK) LTD.**
  27 Old Gloucester Street, London, WC1N 3AX, UK
  Visit us at www.newacademicscience.co.uk • e-mail: info@newacademicscience.co.uk
- **Nairobi** **NEW AGE GOLDEN (EAST AFRICA) LTD.**
  Ground Floor, Westlands Arcade, Chiromo Road (Next to Naivas Supermarket)
  Westlands, Nairobi, KENYA, **Tel.:** 00-254-713848772, 00-254-725700286
  Visit us at www.newagepublishers.com • e-mail: kenya@newagepublishers.com

BRANCHES

- **Bangalore** 37/10, 8th Cross (Near Hanuman Temple), Azad Nagar, Chamarajpet, Bangalore- 560 018
  Tel.: (080) 26756823, **Telefax:** 26756820, **E-mail: bangalore@newagepublishers.com**
- **Chennai** 26, Damodaran Street, T. Nagar, Chennai-600 017, **Tel.:** (044) 24353401
  Telefax: 24351463, **E-mail: chennai@newagepublishers.com**
- **Cochin** CC-39/1016, Carrier Station Road, Ernakulam South, Cochin-682 016
  Tel.: (0484) 2377303, **Telefax:** 4051304, **E-mail: cochin@newagepublishers.com**
- **Guwahati** Hemsen Complex, Mohd. Shah Road, Paltan Bazar, Near Starline Hotel
  Guwahati-781 008, **Tel.:** (0361) 2513881, **Telefax:** 2543669
  E-mail: guwahati@newagepublishers.com
- **Hyderabad** 105, 1st Floor, Madhiray Kaveri Tower, 3-2-19, Azam Jahi Road, Near Kumar Theater
  Nimboliadda Kachiguda, Hyderabad-500 027, **Tel.:** (040) 24652456, **Telefax:** 24652457
  E-mail: hyderabad@newagepublishers.com
- **Kolkata** RDB Chambers (Formerly Lotus Cinema) 106A, 1st Floor, S N Banerjee Road
  Kolkata-700 014, **Tel.:** (033) 22273773, **Telefax:** 22275247
  E-mail: kolkata@newagepublishers.com
- **Lucknow** 16-A, Jopling Road, Lucknow-226 001, **Tel.:** (0522) 2209578, 4045297, **Telefax:** 2204098
  E-mail: lucknow@newagepublishers.com
- **Mumbai** 142C, Victor House, Ground Floor, N.M. Joshi Marg, Lower Parel, Mumbai-400 013
  Tel.: (022) 24927869, **Telefax:** 24915415, **E-mail: mumbai@newagepublishers.com**
- **New Delhi** 22, Golden House, Daryaganj, New Delhi-110 002, **Tel.:** (011) 23262368, 23262370
  Telefax: 43551305, **E-mail: sales@newagepublishers.com**

ISBN: 978-81-224-3884-0
C-16-05-9309

Printed in India at Glorious Printers, Delhi.
Typeset at In-house, Delhi.

NEW AGE INTERNATIONAL (P) LIMITED, PUBLISHERS
7/30 A, Daryaganj, New Delhi-110002
Visit us at **www.newagepublishers.com**

Preface

It has been a long felt ambition, to bring out a book on production of pumps, useful to the pump manufacturers. Pump is supposed to be the HEART of industries. Pumps are used, not only in domestic and agricultural areas, but also in industries. Necessary and essential materials, for the manufacture of pumps, based on the research and experience gained by different authors around the world, are also given in a nutshell. Quick reference can be made by referring the monograms and charts, given.

I will be very much grateful for the valuable assistance given by my wife Smt. Nalini Srinivasan and my two sons, Sri Muthuraman Srinivasan and Sri Jagan Mohan Srinivasan, in producing this manuscript. Also I am very much thankful to M/s New Age International (P) Ltd., Publishers for publishing this book in the correct and commendable form.

Dr. K.M. Srinivasan

Contents

Chapter 3: Testing and Performance Analysis
(Pages: 134–149)

Chapter 4: Systems and Arrangement
(Pages: 150–168)

Chapter 5: Regulation
(Pages: 169–185)

Chapter 6: Effect of Minor Alternations in Pump Parts on Performance
(Pages: 186–196)

Chapter 7: Pump Maintenance
(Pages: 197–223)

1 Pump Classification and Selection

1.1 INTRODUCTION

Newton's law states that "Energy can neither be created nor be destroyed; instead, it can be transformed from one form of energy to another form". A machine is a contrivance, responsible for such conversions. All machines follow the Newton's second law, which states that all forces existing in the world must be converted into the time rate of change of momentum i.e.,

$$\Sigma F \alpha \frac{d(mv)}{dt}$$

where F is the force which is responsible for the energy development, m is the mass of the element, v is the velocity of the mass moving and t is the time. Newton's second law states that all forces (past, present and future) must be converted into a moving force in order to do work i.e., to exist in the world. For example, the human muscle force in the hand must be converted into moving force of the hand, in order to take food, so also to talk, to eat in mouth, etc., The earth and in fact all the planets, in the universe must convert their energy into moving energy in order to rotate around sun, in order to live in the universe.

A wind mill converts the air energy into Mechanical energy with the help of winds. The opposite of this is the ceiling fan we use in our house, where the mechanical energy is converted into air energy.

An electric motor converts electrical energy into mechanical energy. An internal combustion engine converts chemical energy of fuel into mechanical energy. Turbines convert fluid energy into mechanical energy. Hydraulic turbines convert potential energy of water into mechanical energy. Steam turbines convert pressure of steam into mechanical energy and gas turbines convert pressure of fuel gas mixture into mechanical energy. Wind mills convert wind velocity into mechanical energy. Pumps, Fans, Blowers and Compressors convert mechanical energy into fluid energy. Fans convert mechanical energy into kinetic (velocity) energy of air upto 1.0 ata. Blowers convert mechanical energy into pressure energy of air or gas the outlet pressure being 1.2 to 1.5 ata, whereas compressors convert mechanical energy into pressure energy of air or gas above 1.5 ata. Liquid is incompressible. In pumps, mechanical energy is converted into hydraulic energy namely pressure energy, potential energy and kinetic energy of water or any other liquids like milk, distilled water, acids. Pumps are also used for pumping alkaline neutral solutions, cryogenic liquids like Hydrogen, Oxygen, Nitrogen, Ammonia which may be in liquid or in gaseous form, as well as solid-liquid mixtures, liquid-gas mixtures (Two phase flow). Pumping solution may be at a higher or lower temperatures, densities, viscosities and consistencies.

1.2 PUMP CLASSIFICATION

1.2.1 According to Pump Type

Pumps are mainly classified in two ways. In one, pumps are classified according to the pump type (1) Positive displacement pumps and (2) Roto dynamic pumps. In positive displacement pumps, fluid is pushed, whenever pump runs. Fluid movement cannot be stopped, when the pump runs. If stopped, the pressure instantaneously raises to infinity. But when pressure reaches the ultimate strength of the material of the pump, pump breaks suddenly. To avoid this, pressure relief valves are provided as an integral part of the pump. Relief valve immediately opens, when pressure exceeds the safe limit, thereby pump is safe. These pumps are also called static pumps, since the principle of action is purely static. Positive displacement pumps may be reciprocating or rotary or a combination of both types. Under this category comes piston, plunger, ram, lobe, vane, gear, screw, radial

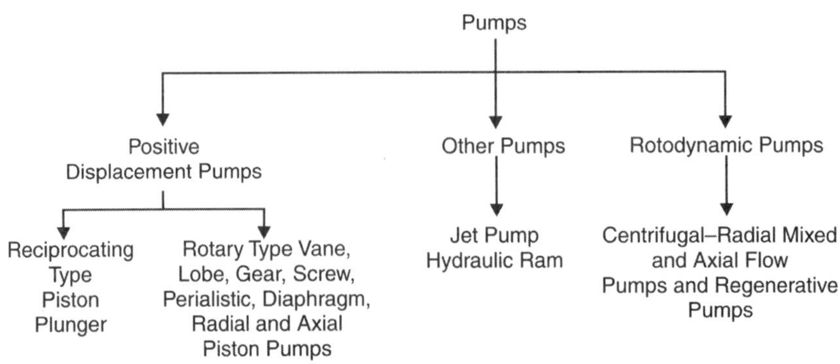

Fig. 1.1: Classifications of pumps

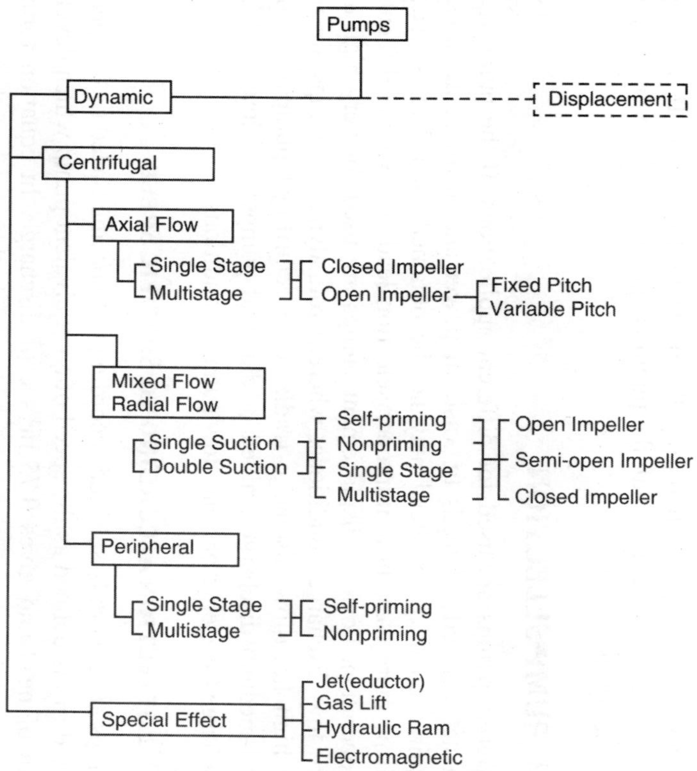

Fig. 1.2: Classification of dynamic pumps

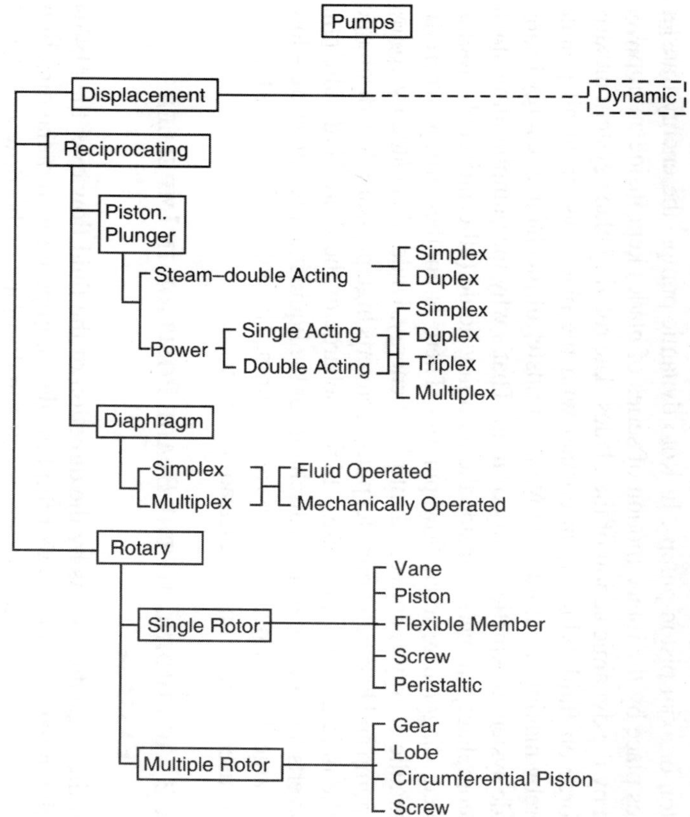

Fig.1.3: Classification of displacement pumps

piston or axial piston pumps. In Roto dynamic pumps, the energy transfer takes place by the rotary motion of series of blades kept inside an impeller and by the dynamic action of the blades. The rotating blade system imparts a force on fluid, which is in contact with the blade system at all points, thereby making the fluid to move *i.e.*, transfer of mechanical energy of the blade system to kinetic energy of fluid. That's why the pumping units must be completely immersed inside the pumping liquid (all centrifugal—radial, mixed, diagonal & axial flow pumps). There is another group, where in hydraulic energy is converted into fluid energy. Under this category comes jet pump and hydraulic ram. Jet pump converts high pressure–low discharge of the fluid into low pressure – high discharge of the fluid. Hydraulic ram converts low pressure – high discharge of the liquid into high pressure – low discharge. Ejector pump is same as jet pump. The only difference is that the air is used as input energy instead of water.

1.2.2 According to the Blade Position in Impeller Passage

Another classification is by the direction of the fluid flow over the blades of the impeller. If the flow of fluid inside the impeller is in radial direction the pump is called radial type centrifugal pump. If the flow is in axial direction the pump is called propeller or Axial flow pump. If the flow is in semi-axial and semi-radial, the pump is called mixed flow or diagonal flow pump (Fig. 1.4).

1.3 PUMP SELECTION

Different pumps are used for different applications. If the pump is not selected correctly, it leads to waste of power and at the same time the required quantity and head may not be obtained. At a later stage this will be more clear how improper selection leads to extreme difficulties in operation. Positive displacement pumps are used for very high head very low discharge conditions, whereas roto dynamic pumps are used for high head-low discharge (radial type centrifugal pumps), medium head-medium discharge (mixed or diagonal pumps) and low head-high discharge (propeller or axial flow pumps) conditions.

1.3.1 Selection according to Specific Speed

Pumps are generally selected according to "Specific Speed (n_s)". Specific Speed (n_s) is defined as the speed of an equivalent pump which develops 1 m of head and gives 0.75 m^3/sec of discharge. In equation form, it is expressed as in metric system where n is the speed of the pump in

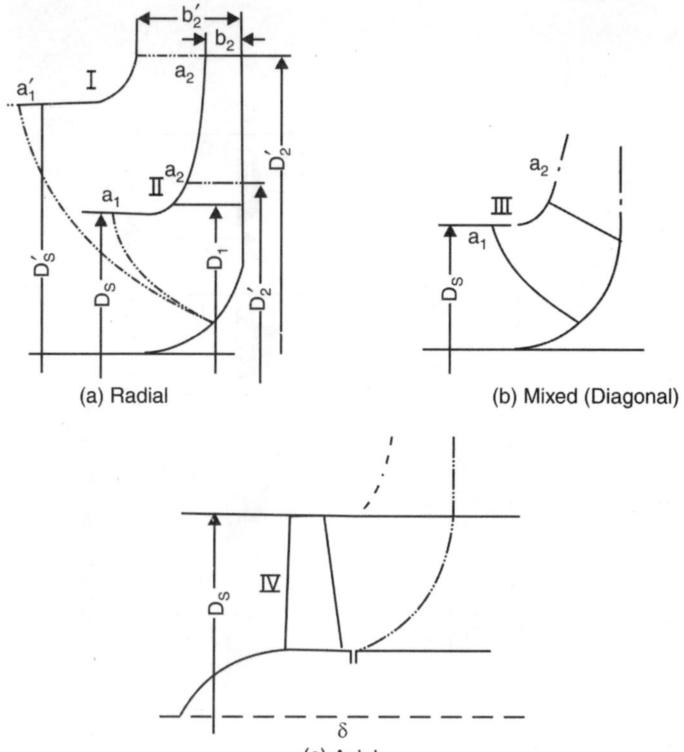

Fig.1.4: Location of blade in impeller passage

$$n_s = \frac{3.65n\sqrt{Q}}{H^{3/4}} \tag{1.1}$$

revolutions per minute (rpm). From the equation, it is evident that $n_s = n$ when $Q = 0.75$ m³/sec and $H = 1$ m.

In British system the Specific Speed $n_s = \dfrac{\text{RPM}\sqrt{\text{GPM}}}{H^{3/4}}$

There is no unit for specific speed. Figures 1.5–1.7 give the selection chart for different types of pump impellers and efficiency according to specific speed. This is the best and correct method of selecting a pump.

1.3.2 Selection according to Head and Discharge

In practice, however, pumps are selected according to the operating conditions, *i.e.*, head and discharge, since this method is quite convenient

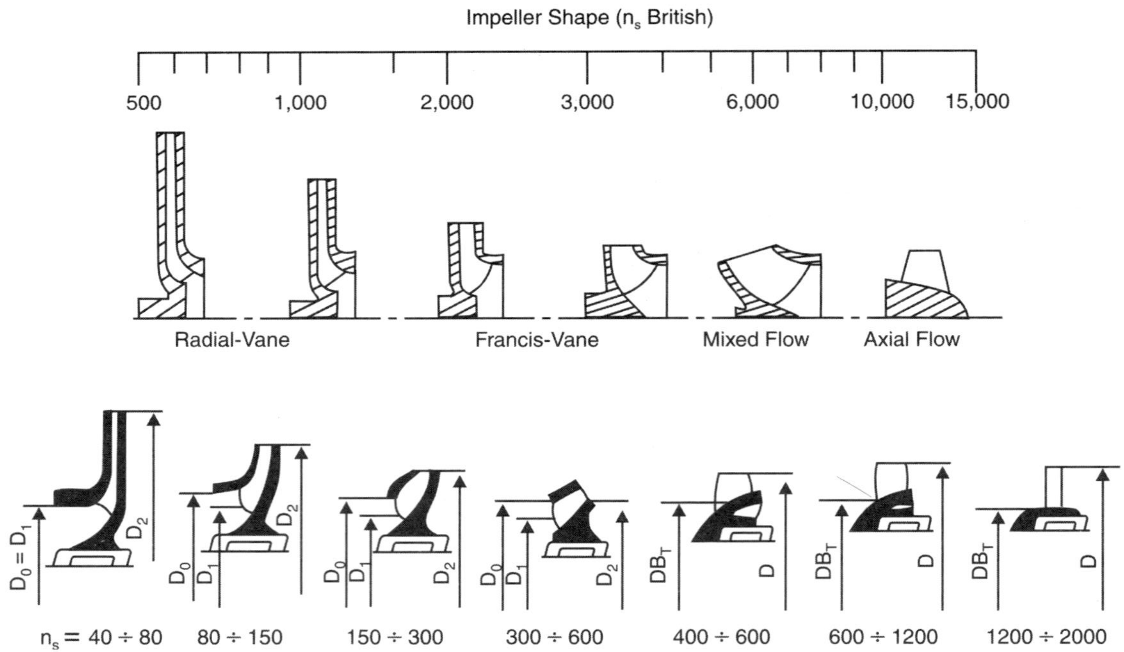

Fig. 1.5: Impeller shapes according to specific speed

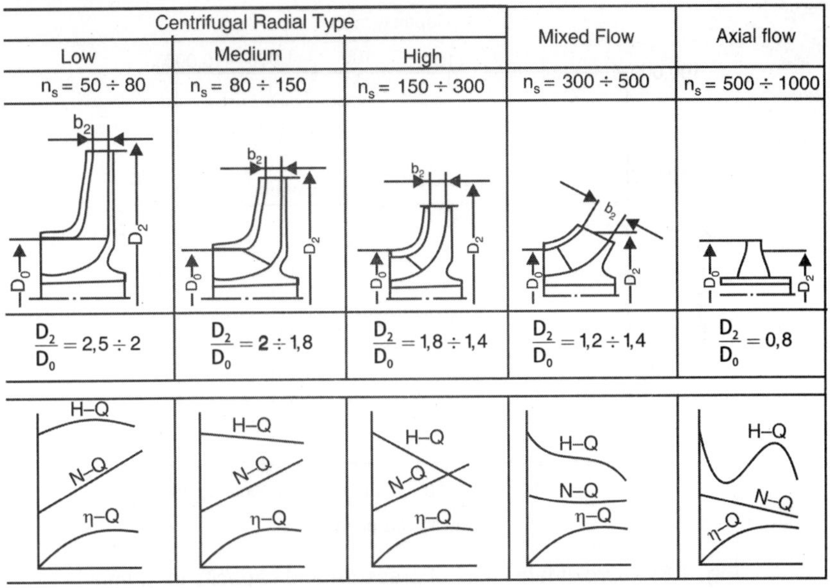

Fig. 1.6: Impeller shapes with their H-Q performance curves

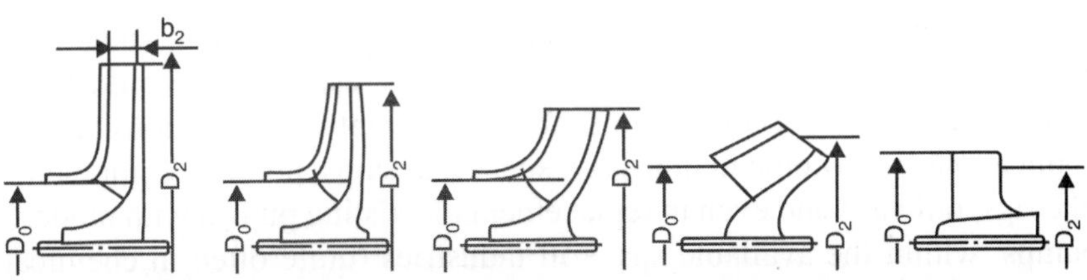

	a	b	c	d	e
n_s	40–80	80–150	150–300	300–600	600–1800
D_2/D_0	~2.5	~ 2	~ 1.8 ÷ 1.4	~ 1.2 ÷ 1.1	~ 0.8 ÷ 0.6

1. n_s = 40 to 20 – Centrifugal low flow (D_2/D_0 = 2.5)
2. n_s = 80 to 140 – Centrifugal normal flow (D_2/D_0 = 2)
3. n_s = 140 to 300 – Centrifugal high flow (D_2/D_0 = 1.0 to 1.4)
4. n_s = 300 to 600 – Diagonal mixed flow (D_2/D_0 = 1.2 to 1.1)
5. n_s = 600 to 1000 - Axial flow d_n/D_0 = 0.3 to 0.6)

Fig. 1.7: Sizes and shapes of impeller according to specific speed (n_s)

for field conditions. Figures 1.8–1.9 and 1.10 give the chart for the selection of pumps according to head and discharge. If pumps are to be operated with liquids other than water, with viscosity, with consistency or with temperature or with density the same chart is used but the parameters of the pump in liquids are converted into equivalent parameters in water.

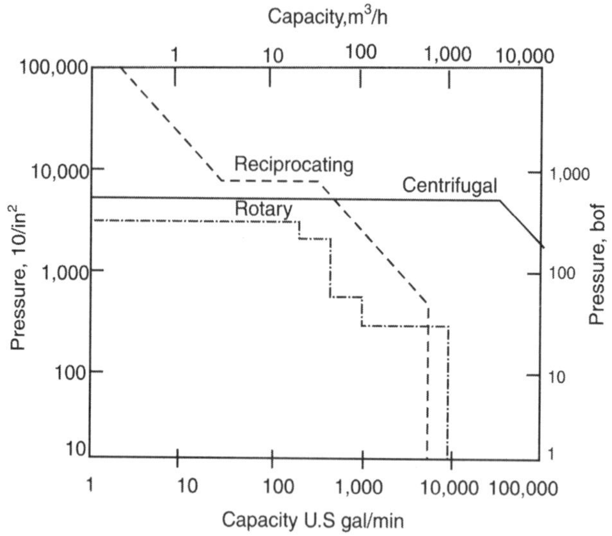

Fig. 1.8: Approximate limits of pressure and capacity for pumps

1.3.3 Special Considerations in Selecting Pumps

Pumps are operated usually at the best efficiency condition, since cost per unit weight of pumping is less. Sometimes, especially in industries, pumps are selected according to space and weight considerations. These applications often come in transport, such as in ships, portable equipments like rigs, railways and even in replacement of existing pumps with modern pumps, within the available space in industries (quite often in chemical industries). In such cases, efficiency is not considered as a major criteria, but the functional application, *i.e.*, non-stop and positive working even at emergency conditions is considered as most important. These change from place to place and installation to installation.

1.4 PUMP PARAMETERS

Independent of the type of pump and its principle of operation, all pumps are specified by the following parameters. (1) Head (2) Discharge (3) Power and (4) Specified speed. Other parameters used for specific reasons are (5) Efficiency and (6) Net Positive Suction Head (NPSH).

1.4.1 Total Head (or) Energy Per Unit Weight of the Liquid

Hydraulic Energy output of the pumps consists of pressure Energy, (p/γ), potential energy (Z), and kinetic energy $(C^2/2g)$. **Bernoulli's equation**

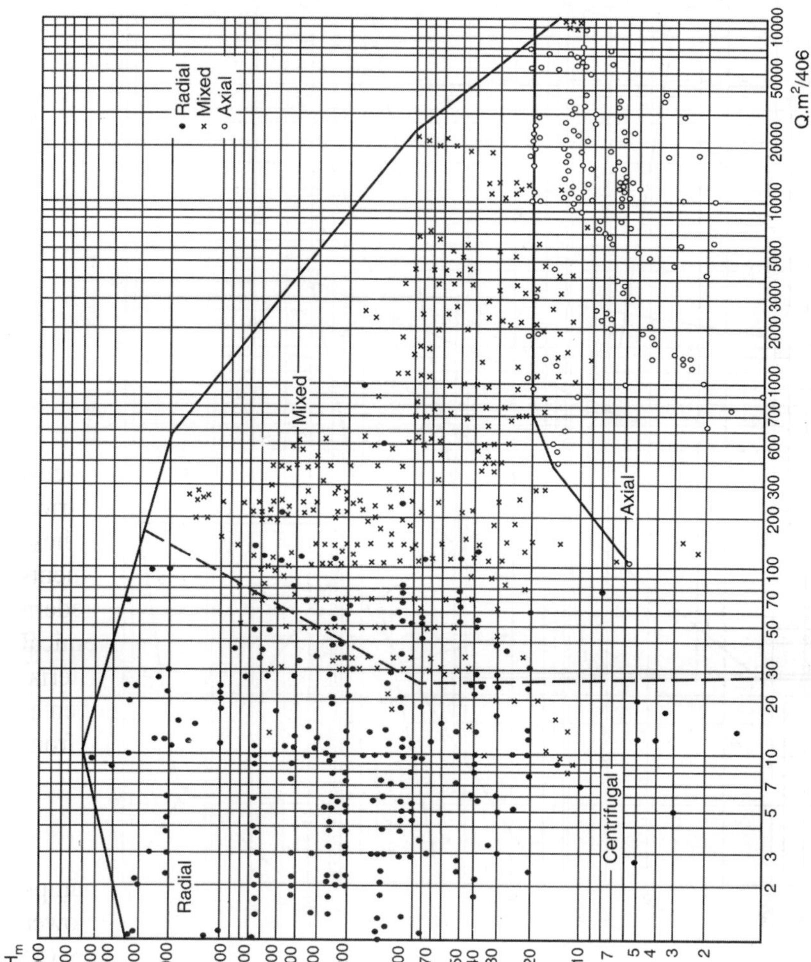

Fig. 1.9: Pump selection according to head and discharge

1. External Gear
2. Sliding Vane
3. Heavy Duty Vane
4. Peripheral
5. Plunger-Axial Piston
6. Sliding Vane
7. Peripheral
8. Flexible Stator
9. Hydraulic-Diaphram
10. Fixed Vane
11. Fixed Vane
12. Flexible Vane
13. Diaphram-Mechanical
14. Fixed Vane
15. Single
16. Eccentric Rotor
17. Single Stage Centrifugal
18. Multistage
19. Flexible Stator
20. Reciprocating Power
21. Multistage Centrifugal
22. Reciprocating (Direct Acting)
23. 2 Stage Centrifugal
24. Internal Gear
25. Sumersible
26. Single Stage Centrifugal
27. Mixed Flow
28. Axial Flow
29. Hydraulic Diaphram

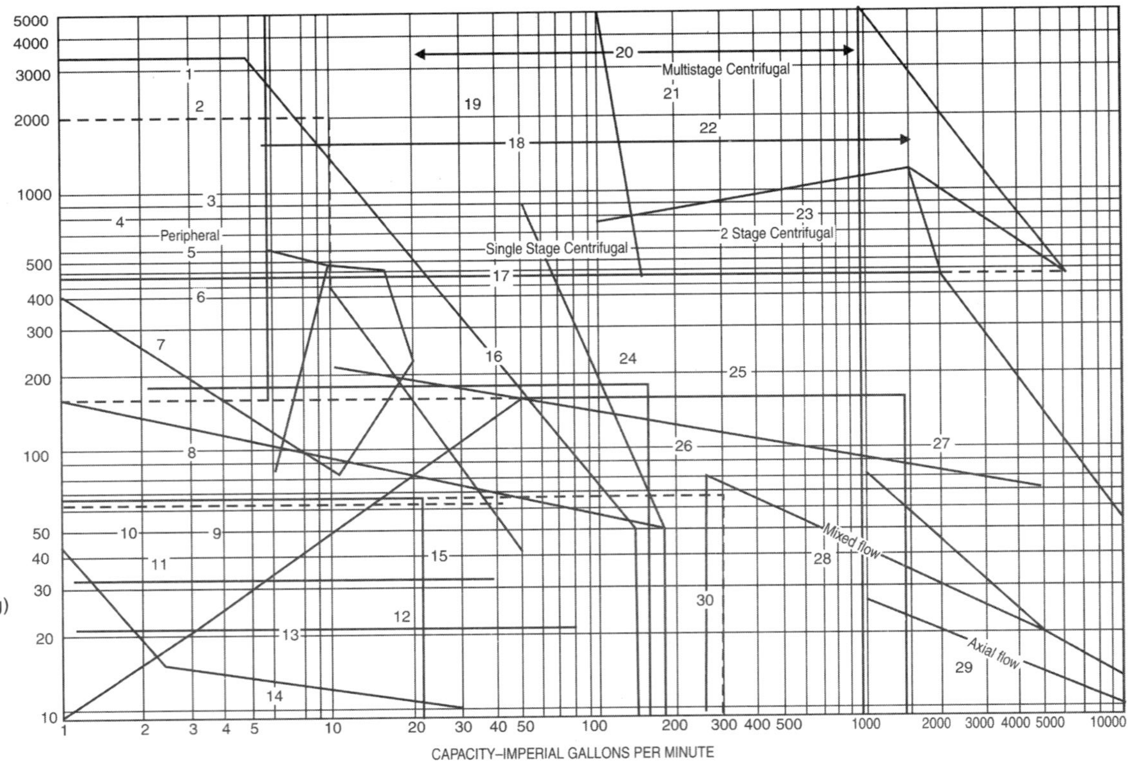

Fig. 1.10: Pump selection according to head discharge

states that "The sum of the pressure energy, potential energy and kinetic energy along a stream line is always constant for a flowing fluid"

$$i.e., \frac{p}{\gamma} + Z + \frac{C^2}{2g} = \text{Constant} \qquad (1.2)$$

at different levels, then the pump output will be in the form of potential energy.

where p = Pressure (kGf/cm^2, N/m^2, Pa, atm, ata) = γH

γ = Specific weight of the flowing fluid (kGf/cm^3, N/m^3)

= ρg (where ρ – density (kG/m^3), g – gravitational acceleration)

Z = Elevation or height above a reference line (m)

C = Velocity of flow (m/sec)

H = Head (m)

(Each term in Bernoulli's equation represents energy per unit weight of fluid) (kGf.m/kGf) or (N.m/N) or (m). The constant is called (H) the Total head of a Pump.

Referring to equation (1.2), it is evident that the energy at the outlet of the pump (at the outlet of the delivery flange), contains all three energies (*i.e.,* pressure, potential and kinetic energies). However pump output may be in any one form of the three energies or may possess any two of the three energies or may be with all three forms of energies. In agricultural and domestic pumps, the output energy is mostly in the form of potential energy *i.e.,* raising water from a lower level to a higher level. In fire fighting and sprayer pumps, the output energy is purely in the form of kinetic energy. *i.e.*, to move water at high velocity. In process pumps, the output energy is mostly in the form of pressure energy, when the suction and delivery tanks are at the same level. If these two tanks are at different pressure energy, located at different places and at different levels, then the pump has to handle all the three energies *i.e.*, pressure, and potential energies.

Apart from this, pumps may have to handle liquids with different temperatures, densities, viscosities, as well as consistencies. Pumping liquid may be either solid-liquid, or liquid-gas or liquid-liquid mixture. All these aspects will be discussed in the following paragraphs.

Pressure head or head of a pump (H) (in meters of liquid column MLC) is defined as increase in the energy per unit weight of fluid passing through the pump, *i.e.*, difference between the energy at the outlet and at the inlet of the pump.

Referring to Fig. 1.11, the theoretical (or) ideal energy will be

Hydraulic energy at outlet $E_2 = \frac{p_2}{\gamma} + Z_2 + \frac{C_2^2}{2g}$ [1.3(*a*)]

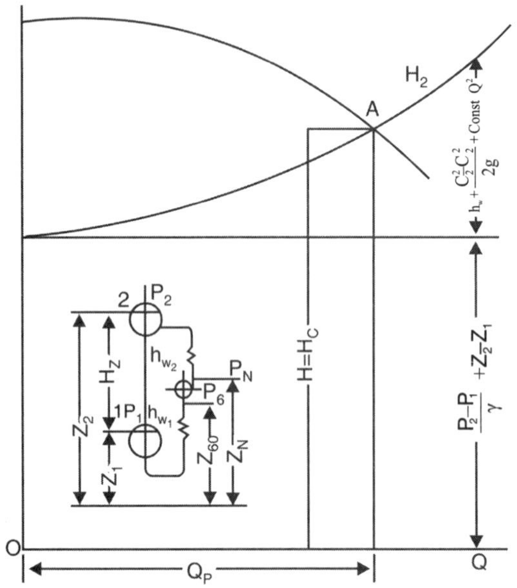

Fig.1.11: Head measurement in a pump with a system

Hydraulic energy at inlet $E_1 = \frac{p_1}{\gamma} + Z_1 + \frac{C_1^2}{2g}$ [1.3(*b*)]

Net energy transferred from mechanical to Hydraulic through impeller will be the Total head (*H*) in meters = $E_2 - E_1$.

1.4.2 Quantity of Flow or Discharge

Quantity of flow or discharge of a pump is defined as the amount of fluid given by the pump in unit time. It can also be explained as the quantity of liquid passing through the pump in unit time. Depending upon the conditions of service, the quantity of flow is measured in volume or weight per unit time. Normal units adopted are cubic meter/hour (m^3/hr); cubic meter/second (m^3/sec); liters/minute (Lpm) liters/second (Lps) in volumetric system and tons/hour (T/hr); kilogram/second (kG/sec) etc., in gravimetric (weight) system. The relation between volumetric and gravimetric system is given by

Weight (W) = Specific weight (γ_w) × Volume (Q)

The value of specific weight (γ_w) for water = 1000 kGf/m³. For liquids other than water, specific weight of liquid $\gamma_s = \gamma_w S_l$, where S_l is the specific gravity of liquid.

1.4.2.1 Flow measurement

(1) Gravimetric Method

A collecting tank of 1 m length (L) × 1 m width (W) × 1¼ m height (H) is placed on a weighing balance. The delivery of water from the pump is directed to the tank so that water is collected in the tank. A gate valve is fitted at the bottom of the tank, in order to empty the tank, whenever necessary. At normal conditions, the water from the collecting tank will be directed to suction sump, by opening the gate valve. To start with, the pointer in the weighing balance dial is set for one particular reading (weight (W_1)). The bottom gate valve in the collecting tank is now closed completely. Water gradually gets collected in the collecting tank and the weight of water collected in the tank gradually increases. This is shown by the gradual increase in the weighing balance dial reading. A stop watch is started as soon as the weight of water collected in the collecting tank, reaches the set weight reading in the dial ($t_1 = 0$). The set weight reading in the dial now is set to a higher reading (W_2). After specific time, the weight of water collected will reach the newly set reading (W_2). The stop watch is stopped as soon as the weight of water collected in the tank reaches the newly set reading (t_2). Thus the time taken ($t_2 - t_1$) to collect specific amount of weight of water from the pump ($W_2 - W_1$) in the collecting tank is known. From these readings, the weight of water from the pump for one second can be calculated. The bottom gate valve is opened now, immediately, so that water does not overflow from the collecting tank.

Net weight of water collected (W) = ($W_2 - W_1$) in kGf

Time taken to collect this weight of water (t) = $t_2 - t_1$ in seconds

Weight Flow rate through the pump = (W/t) in kGf/sec

This is the most accurate method. It includes variation in density also along with quantity. It gives an error of 0.5 %.

(2) Volumetric Method

A collecting tank of 1 m length (L) × 1 m width (W) × 1¼ m height (H), fitted with a glass tube and a graduated scale, at the outside surface of the tank, in order to measure the height of water collected in the tank, is used to collect, the delivery of water from the pump.

A gate valve is fitted at the bottom of the tank, in order to empty the tank, whenever necessary. At normal conditions, the water from the collecting tank will be directed to suction sump, by opening the gate valve. To start with, the gate valve in the collecting tank is closed completely. Water level in the tank gradually rises. This is shown by the corresponding rise in level in the glass tube. A stop watch is started at any water level (H_1) in the collecting tank ($t_1 = 0$). For a specific rise in water level in the collecting tank $H_2 - H_1 = 10$ to 15 cms), the stop watch is stopped and the time is noted (t_2). Thus, the time taken ($t_2 - t_1$) to collect specific amount of water from the pump in the collecting tank is known. From these readings, the amount of water delivered by the pump for one second can be calculated. The bottom gate valve is opened now, immediately, so that water does not overflow from the collecting tank.

Area of the collecting tank = length (L) × width (W) in m^2

Rise of water level in the tank (H) = $H_2 - H_1$

Hence volume of water collected (V) = $L \times W \times N$

Time taken to collect this volume of water (t) = ($t_2 - t_1$) in seconds

Volume flow rate through the pump = (V/t) in m^3/sec

This is the second accurate method. It only considers variation in volume. The density variation is neglected. It gives an error of 1% to 1.5%.

(3) Flow Instruments

As per Bernoulli's equation, a change in velocity along a closed conduit induces a change in pressure for the same level ($Z_1 = Z_2$). This pressure drop is measured by a manometer, to determine the flow in any closed conduits. Change in velocity is done by a change in area of cross section of the conduit.

Venturi meters, Orifice meters, and Nozzles are the three meters, used to measure the flow in closed conduits. Orifice meters are also used in tanks to measure the flow into or from a tank. Venturi meters are used mostly for incompressible fluid flow (water and oil) measurement. Orifice meters and Nozzles are used for both incompressible (water and oil) and compressible fluid flow (air and gas) measurement. All these three meters have similar construction. The main diameter at the entrance of the meter is gradually reduced to a throat diameter, in venture meters and Nozzles, whereas there will be a sudden reduction from main diameter to throat diameter in orifice meters. These meters are connected in the pipe line with 4d to 6d (d is the pipe diameter) straight and free upstream and downstream length, in order to ensure uniform flow at the meters. Then only the readings taken from these meters will be correct. A manometer is connected between the main diameter and the throat diameter, in order to measure pressure difference.

Quantity of flow through the meters $Q = C_D \frac{a_1 a_2}{\sqrt{a_1^2 - a_2^2}} \sqrt{2gH}$, in m^3/sec

Where, C_D is the coefficient of discharge, which is a constant, and has a value of

0.96 to 0.98 for Venture meters,

0.85 to 0.88 for Nozzles and

0.62 to 0.64 for Orifice meters

a_1 is the area at the main diameter (d_1) of the meter, which is equal to pipe diameter, $a_1 = \pi \frac{d_1^2}{4}$ in m^2

a_2 is the area at the throat diameter (d_2), which is given by the manufacturer, normally $d_2 = 0.5\ d_1$, $a_2 = \pi \frac{d_2^2}{4}$ in m^2

g is the gravitational constant = 9.81 m/sec^2,

H is the pressure drop across the main and the throat diameters of the meter, which is measured by the mercury manometer, connected to the meter in m.

This method is commonly used in industries. It gives an error of 3% to 5%.

(4) Flow Meters

Flow meters are directly connected in pipe lines, like flow instruments. They give the flow directly. A rotor with screw shaped blades will be rotating across the flow, inside pipes, whenever flow takes place. This rotation is connected to a dial through gears. The dial is graduated in flow rate. The rotation of rotor blades is proportional to flow rate. The flow meter will be calibrated after manufacture with a standard flow measuring device like volumetric method. These meters are used in industries, houses and other places where the flow rate is the only requirement. These meters give an error of approximately 10%.

(5) Field Testing

$$v = \frac{X}{t} \text{ or } t = \frac{X}{v} \qquad [1.4(a)]$$

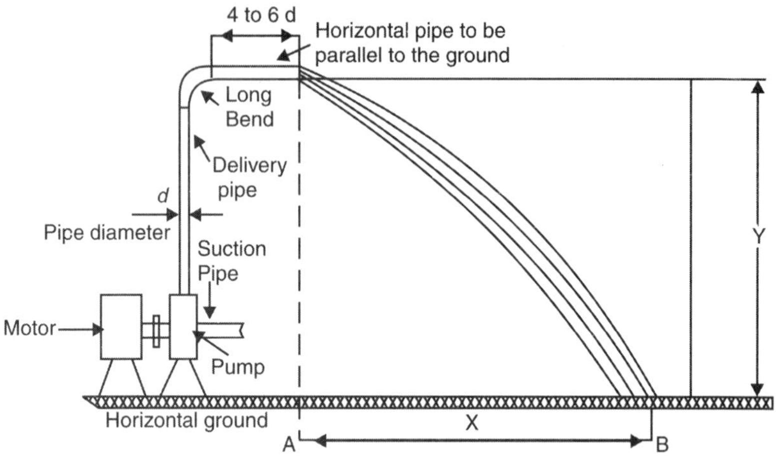

Fig. 1.12: Flow measurement in field

$$Y = \frac{1}{2} g\, t^2 \text{ or } \frac{1}{2} g \frac{X^2}{v^2} \qquad [1.4(b)]$$

$$v = \sqrt{\frac{gX}{2Y}} = 2.215 \frac{X}{\sqrt{Y}} \qquad [1.4(c)]$$

The formula used $v = 2.215 \frac{X}{\sqrt{Y}}$, where v is the velocity in m/sec.

X is the horizontal distance of fall of water from the delivery pipe in meters (m).

Y is the vertical distance between the pipe and the ground in meters (m).

Table 1.1: Velocity of water coming out of the delivery pipe, (v), in m/sec

***X* m** ***Y* m**	**5** **m**	**6** **m**	**7** **m**	**8** **m**	**9** **m**	**10** **m**	**11** **m**	**12** **m**	**13** **m**
2 ½ m	7	8.41	9.81	11.21	12.61	14.00	15.41	16.81	18.21
3 m	6.39	7.87	8.95	10.23	11.51	12.70	14.07	15.35	16.62
3 ½ m	5.92	7.10	8.29	9.47	10.66	11.84	13.02	14.21	15.39

$A = \frac{\pi d^2}{4}$ where A is in sq. meters = A in sq. inches × 6.4516 /10000

Table 1.2: Area of pipe for the given diameter

Diameter in inches (*d*)	Area in sq. inches (*A*)	Area in sq. meters (*A*)
1½ inch	1.7672 sq. inch	1.14×10^{-3}
2 inch	3.1415 sq. inch	2.02×10^{-3}
2 ½ inch	4.9087 sq. inch	3.17×10^{-3}
3 inch	7.0686 sq. inch	4.56×10^{-3}
4 inch	12.5664 sq. inch	8.11×10^{-3}
6 inch	28.2743 sq. inch	18.20×10^{-3}

$$Q \text{ (lit/sec)} = A \text{ (sq. meters)} \times v \text{ (m/sec)} \times 1000, \text{ where } A = \frac{\pi d^2}{4}$$

Table.1.3: Quantity of flow through the installed delivery pipe, (*Q*),in lit/sec

(*v*) m/sec (*d*) inch	4 m/sec	5 m/sec	6 m/sec	7 m/sec	8 m/sec	9 m/sec	10 m/sec
1½	4.56	5.70	6.86	7.98	9.12	10.26	11.40
2	8.08	10.10	12.12	14.14	16.16	18.18	20.20
2½	12.67	15.14	19.00	22.10	25.34	28.50	31.67
3	18.24	22.80	27.36	31.92	36.48	41.04	45.60
4	32.43	40.54	48.65	56.76	64.16	72.97	81.08
6	72.80	91.00	109.20	127.40	145.60	163.80	182.00

1.4.3 Power

Power spent by the pump is measured in kilowatt (kW). Useful power depends upon the head and discharge of the pump. The product of head in metres and weight of the liquid pumped per unit time in kGf/sec, is the energy (useful) spent (kGf.m/sec). Since 1 kW = 1 kGf.m/102 sec.

Useful power of the pump (P_o) will be $\frac{WH}{102}$ kW.

In SI unit specific weight of water $\gamma = 9.81 \times 1000$ N/m^3 = 1000 kGf/m^3 and also

1 kW = 1 N.m/1000 sec.

Hence useful power of the pump $N_0 = \frac{WH}{1000} = \frac{\gamma QH}{1000}$

where γ is in N/m^3; Q in m^3/sec, H in m and W in N/sec.

1.4.4 Losses and Efficiency

Theoretically, all the energy supplied to the pump by the prime mover, in the form of mechanical energy, is converted into fluid energy. Owing to manufacturing inaccuracies and entirely different flow conditions prevailing in pump, entire energy input (mechanical energy) is not converted into fluid energy. Referring to Figs. 2.19 and 2.20, 100% mechanical energy supplied at the coupling side of the pump by the prime mover is reduced, due to energy absorption in bearings, stuffing box, disc friction. Hence, the energy input at the impeller will be less than the energy input at the pump coupling. Due to surface roughness inside impeller and due to the leakage flow through clearance, there will be further reduction in the energy input to the impeller. Hence, the energy output from the pump is less than the energy input to the pump. The difference between energy input and energy output of the pump is called losses in pump. The ratio of energy usefully utilized for work to the energy supplied is called efficiency. In other words, efficiency is the percentage of useful energy available from the input energy of the machine in doing work.

Three kinds of losses prevail in fluid machines namely, (1) Hydraulic losses (2) Volumetric losses and (3) Mechanical losses. The sum of all losses will be the total loss. Overall efficiency is the product of hydraulic efficiency, volumetric efficiency and mechanical efficiency.

1.4.4.1 Hydraulic loss (ΔH) and hydraulic efficiency (η_h)

Due to surface roughness at the inner side of the impeller, through which the fluid passes, losses due to friction and losses due to secondary flow, and hence energy losses occur. Actual head developed (H_a) will be less than the theoretical head (H_{th}) by the amount $\Delta H = H_{th} - H_a$.

ΔH is called the hydraulic loss. Hydraulic efficiency (η_h) is the ratio between actual head to the theoretical head.

Hydraulic loss, $\Delta H = H_{th} - H_a$

Hydraulic efficiency, $\eta_h = \dfrac{H_a}{H_{th}} = \dfrac{H_{th} - \Delta H}{H_{th}} = \dfrac{H_a}{H_a + \Delta H} = 1 - \dfrac{\Delta H}{H_{th}}$

$$\Delta H = (1 - \eta_h)\, H_{th} \qquad (1.5)$$

1.4.4.2 Volumetric loss (ΔQ) and volumetric efficiency (η_v)

In order that the impeller can rotate freely inside the stationary casing, proper clearance is provided at the front and at the rear side of the impeller at wearing rings. Due to pressure difference between impeller outlet and impeller inlet at the front side of the impeller as well as the pressure difference between impeller outlet and slightly higher than atmospheric pressure at the stuffing box, part of fluid coming out of the impeller leaks through the clearances on both sides of the impeller. As a result the quantity coming out of the pump, the actual quantity (Q_a) will be less than the quantity passing through the impeller *i.e.*, theoretical quantity (Q_{th}) will be reduced by the amount of leakage quantity passing through the clearances (ΔQ) *i.e.*, $\Delta Q = Q_{th} - Q_a$. Volumetric efficiency (η_v) is the ratio between actual quantity and theoretical quantity

Volumetric Losses $\Delta Q = Q_{th} - Q_a$

Volumetric Efficiency $\eta_h = \dfrac{Q_a}{Q_{th}} = \dfrac{Q_{th} - \Delta Q}{Q_{th}} = \dfrac{Q_a}{Q_a + \Delta Q} = 1 - \dfrac{\Delta Q}{Q_{th}}$ (1.6)

$$\Delta Q = (1 - \eta_h) Q_{th}$$

1.4.4.3 Mechanical loss and mechanical efficiency (η_m)

Energy loss in ball, roller or thrust bearings (ΔN_B), in bush bearings at stuffing box or in mechanical seal portion (ΔN_s), and the disc friction losses (ΔN_D) due to the impeller rotation inside the volute casing, which is filled with fluid are classified as mechanical losses (ΔN). The energy received at the impeller side of the shaft *i.e.*, actual power (N_i) for energy conversion into fluid energy will be less than the energy supplied at the coupling side by the prime mover, *i.e.*, theoretical power. (N_{th}), *i.e.*, $\Delta N = N_{th} - N_i$. The ratio between actual power (N_i) and the theoretical power (N_{th}) is the mechanical efficiency (η_m) *i.e.*,

Total Mechanical Losses

$\Delta N = \Delta N_B + \Delta N_S + \Delta N_D$

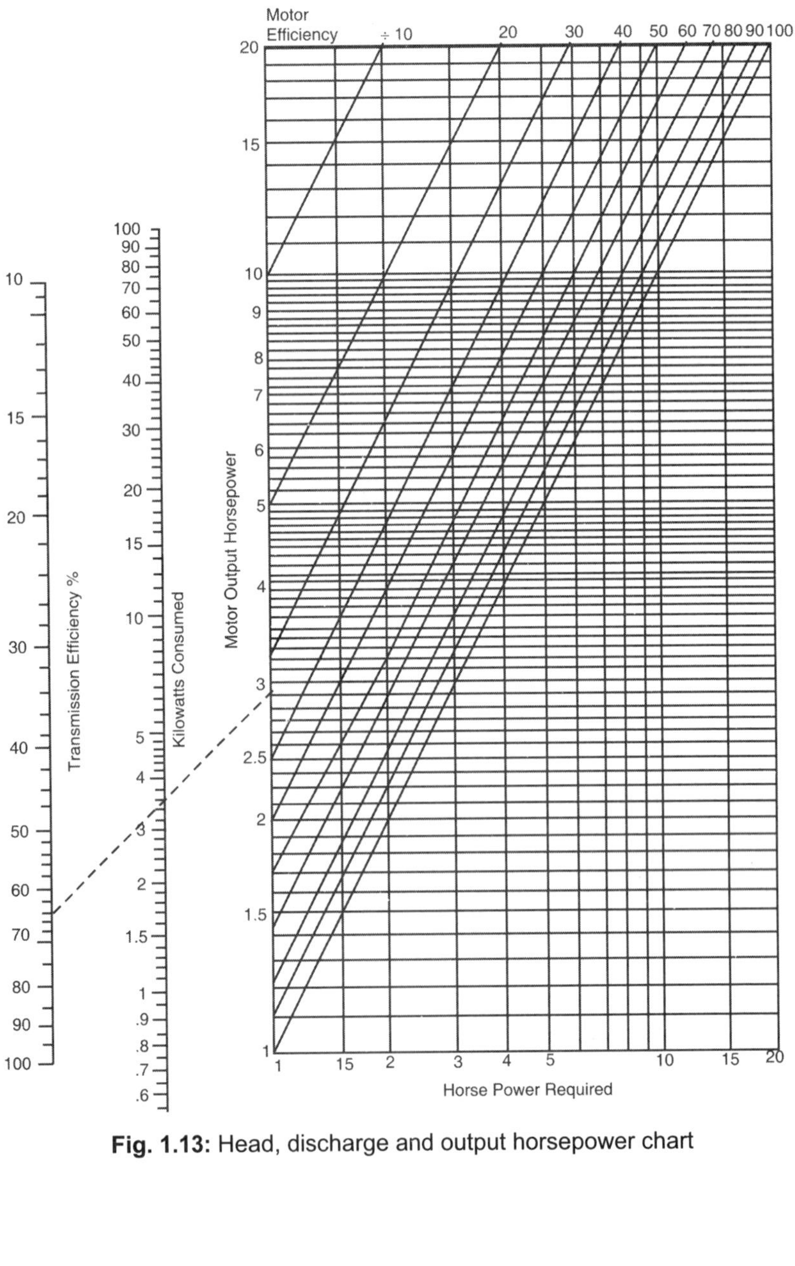

Fig. 1.13: Head, discharge and output horsepower chart

$\Delta N = N_{th} - N_{i.}$ Where N_i = power available at the impeller end of the shaft,

$$\text{Mechanical Efficiency } \eta_m = \frac{N_i}{N_{th}} = \frac{N_{th} - \Delta N}{N_{th}} = \frac{N_i}{N_i + \Delta N} = 1 - \frac{\Delta N}{N_{th}} \quad (1.7)$$

$$\Delta N = (1 - \Delta\eta)\, N_{th}$$

1.4.4.4 Total losses and overall efficiency (η)

Total losses = Hydraulic loss + Volumetric loss + Mechanical loss

$$= \Delta H + \Delta Q + \Delta N.$$

Since $\eta_v = \dfrac{Q_a}{Q_{th}}$, and $\eta_h = \dfrac{H_a}{H_{th}}$

Output energy $(N_0) = \gamma Q_a . H_a = \gamma Q_{th} . \eta_v . H_{th} . \eta_h$

But $N_i = \gamma Q_{th}\, H_{th}$

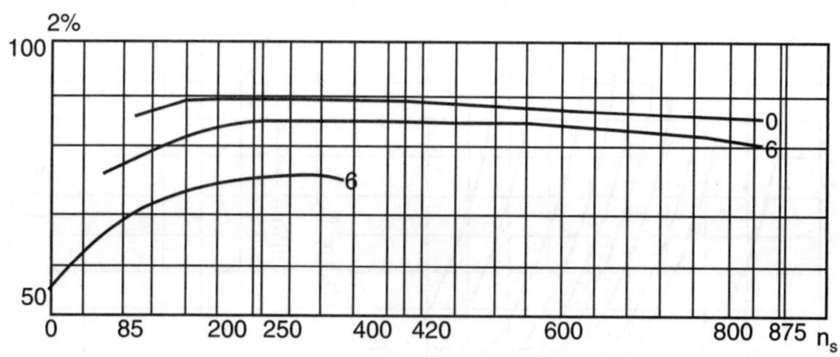

(a) D_0 > 200 mm,
(b) D_0 = 200 to 50 mm
(c) D_0 = 50 to 10 mm

Fig. 1.14: Efficiency vs specific speed for different sizes $\eta = f(n_s, D_0)$

Since $\eta_m = \dfrac{N_i}{N_{th}}$, $N_0 = N_i \eta_v\, \eta_h = N_{th}\, \eta_m .\, \eta_v .\, \eta_h .$

$$\text{Overall efficiency, } \eta = \frac{N_0}{N_{th}} = \eta_m \eta_v \eta_h$$

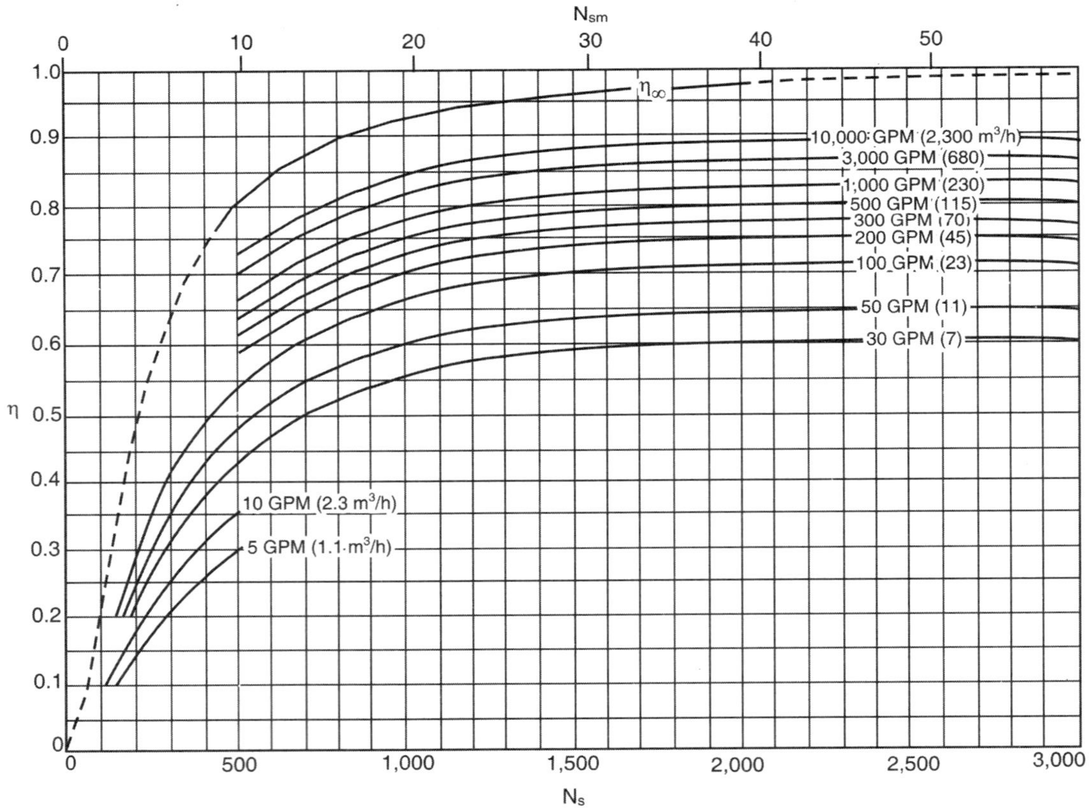

Fig. 1.15: Efficiency as a function of specific speed and capacity

1.4.5 Suction Conditions, Limitations and Applications

If a vacuum tube of sufficient length is inserted into a tub containing mercury, the mercury in the tube will stand to a height of 760 mm. This is because of atmospheric pressure pressing the mercury at the outside area of the tube, making the mercury to raise into the tube space. Since the tube is under vacuum, the mercury raises to a height equal to the outside atmospheric pressure. If instead of mercury, water is used, the height would have been 10.336 m inside the tube. If air is used the height will be 10000 m. It means that a height of 10000 m of air column above earth prevails and that exerts a pressure on us, and on earth, which is called atmospheric pressure. Under static condition, under vacuum and at normal temperature, water can rise to a maximum height, equal to that of atmospheric pressure of the surroundings or what is called field conditions. In centrifugal pumps, vacuum is created at the eye of the impeller, which makes the liquid to pass through the

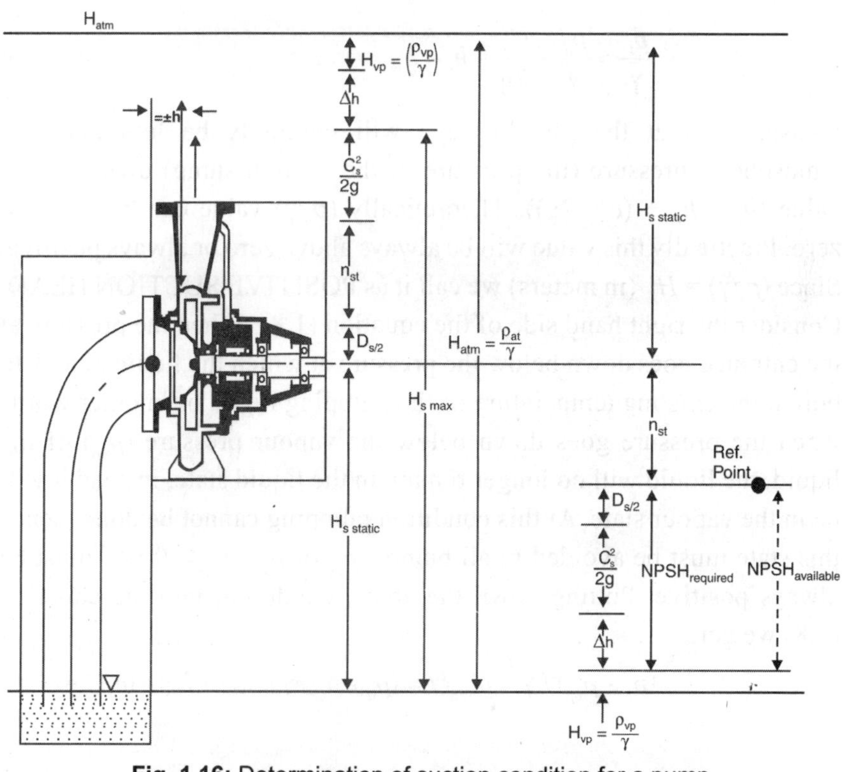

Fig. 1.16: Determination of suction condition for a pump

suction pipe from the suction sump. Since water in the centrifugal pump is always moving, actual suction effect of the pump will be lower than that of atmospheric pressure *i.e.*, less than 10.336 m. This value depends upon the quantity of flow (or velocity of flow) of liquid through the pipe, the temperature of the pumping solution, height of pump above the water level in the sump and also depends upon the suction sump pressure. In practice, the suction condition of a pump is measured by means of a term NPSH (Net Positive Suction Head).

1.4.5.1 Net positive suction head (NPSH)

Applying Bernoulli's theorem, and equating the energy at the suction sump and at the entrance of the impeller, taking reference as the line passing through the water level in sump, Fig. 1.16,

$$\frac{p_a}{\gamma} = \frac{p_s}{\gamma} + \frac{C_s^2}{2g} + h_s + h_{fs} \qquad [1.8(a)]$$

$$\frac{p_s}{\gamma} = \frac{p_a}{\gamma} - \frac{C_s^2}{2g} - h_s - h_{fs} \qquad [1.8(b)]$$

Pressure (p_s) at the impeller eye will definitely be less than the atmospheric pressure (the pressure of the suction sump) owing to the value ($h_s + h_{fs} + (c_s^2/2g)$). Theoretically ($p_s/\gamma$) value can be equal to zero. Practically this value will be always above zero or always positive. Since (p_s/γ) = H_{sv} (in meters) we call it as POSITIVE SUCTION HEAD. Consider the right hand side of the equation (1.8). When the pressure at the entrance goes down below the pressure at which the liquid begins to boil at the existing temperature of the pumping liquid or in other words when the pressure goes down below the vapour pressure (p_{vp}) of the liquid, the liquid will no longer remain in the liquid state, instead it will be in the vapour state. At this condition pumping cannot be done, hence this state must be avoided in all pumps. p_{vp} or $p_s - p_{vp} > 0$ *i.e.*, must be always positive. Putting down the above condition, into the equation (1.8) we get,

$$(p_s - p_{vp}) / \gamma = p_{\min}/\gamma = (p_a - p_{vp})/\gamma - (h_s + h_{fs} + (c_s^2/2g))$$

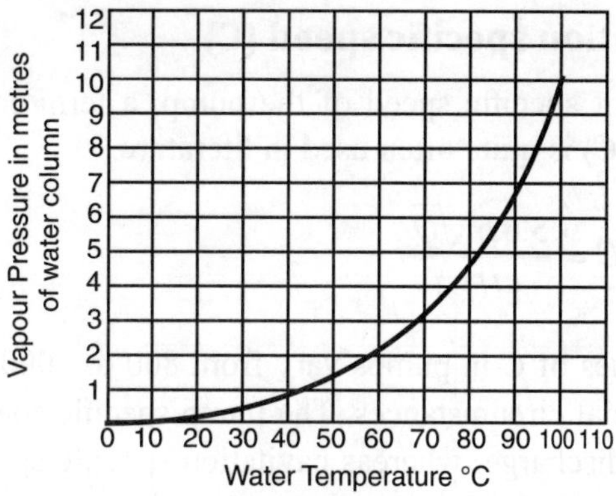

Fig. 1.17(a): Vapour pressure—Temperature for water

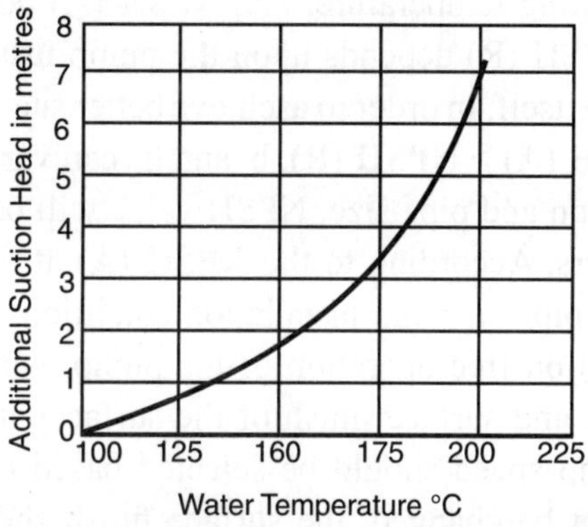

Fig. 1.17(b): Additional suction head—Temperature for water

or $H_{sv} = (H_a - H_{vp}) - (h_s + h_{fs} + (c_s^2/2g))$ or NPSH = NPSH(A) – NPSH(R). In order that H_{sv} should be above zero or always positive, **NPSH** available **(NPSH (A)) > NPSH** required **(NPSH (R)).**

1.4.5.2 Thoma's constant (σ)

In literature, non-dimensional parameter such as Cavitation Coefficient or dynamic depression, is used. This coefficient is called **THOMA'S Constant**

$$\text{Thoma's Constant } (\sigma) = \frac{\text{NPSH or } H_{SV} \text{ or } \Delta H}{H} \tag{1.9}$$

where H is the total head of the pump.

1.4.5.3 Cavitation specific speed (C)

Similar to that of specific speed of the pump, a term called Cavitation Specific Speed (*C*) is quite often used in literature.

$$C = \frac{5.62n\sqrt{Q}}{\left(H_{sv}^{3/4}\right)} \tag{1.10}$$

Normal values of *C* in pumps vary from 800 to 1000 and even up to 2000 under special circumstances. The pump specific speed (n_s) depends upon head and discharge, whereas cavitation specific speed (*C*) depends upon, H_a, the atmospheric pressure or the sump pressure and H_{vp} the vapour pressure depends on the surrounding condition as well as the liquid property and operating temperature. *i.e.*, NPSH (A) does not depend on pump, whereas NPSH (R) depends upon the pump installation condition as well as the pump itself. In order to aschieve better suction characteristics it is essential **NPSH (A) > NPSH (R)**. h_s and h_{fs} can vary according to the installation condition and pipe size. NPSH value will be always given by pump manufacturers. According to the **NPSH (A)** at the site conditions, the proper suction pipe size and installation conditions are established in order to get cavitation free operation of the pump, which depends upon speed of the pump and surface finish of the surfaces in contact with the fluid. Strictly, pump speed should be selected based on the value of *C*. If the speed cannot be changed, the surface finish should be improved, which reduces the friction losses. This, in turn, improves the NPSH value. A few graphs are given in the following pages for reference purpose. These graphs will enable the designer to select correctly the pump and its parameters.

1.5 EFFECT OF OTHER PARAMETERS ON PUMP PERFORMANCE

1.5.1 Effect of Temperature

Increase in temperature reduces density (ρ), viscosity, (μ or ν) increases vapour pressure (p_{vp} or H_{vp}) of the pumping liquid and even changes the dimensions of pump parts like shaft, bushes, sleeves of stuffing box (or mechanical seal) wearing rings of impeller and casing, bearing housing etc., due to thermal expansion of the material of pump parts. Pumps, to be operated for high temperature must be manufactured with extra allowance, equal to the thermal expansion of the materials of pump, for the running temperature, over and above the regular manufacturing tolerances and allowances. These pumps should not be started at the normal, atmospheric temperature; instead the pumps are either heated or filled with the pumping liquid, so that the pump attains the operating temperature, before starting. Increase in temperature increases vapour pressure. This reduces the suction conditions, increases NPSH or H_{sv}, with the result, pumps will be operating under cavitation. The material of the pump will be eroded gradually. Pumps produce noise and vibration. Under such conditions pump must either be lowered down or suction tank must be raised, in order to reduce the static suction lift. Sometimes, the tank may be kept above pump level *i.e.,* negative suction lift ($-h_s$) will become positive suction head ($+h_s$). Also the suction tank pressure should be increased. The suction pipe size may also have be increased.

Viscosity of liquids decreases when temperature is increased. Standard charts are available for the determination of correct viscosity at the operating temperature Viscosity conversion charts are used for the determination of the pump parameters.

1.5.2 Effect of Density

Change in density affects the total head of the pump and correspondingly power required for the pump. Pump delivery pressure at the pump outlet will be always same, whatever may be the density

Since $p = \gamma H$ and $\gamma = \rho g$

$p_L = \gamma_L H_L$ and $p_w = \gamma_w H_w$

$\gamma_L = S\gamma w$

γ = Specific weight = $S\gamma_w$ (kGf/m^3 or N/m^3), ρ - density (kG/m^3),

g = Acceleration due to gravity (m/sec^2), S-Specific gravity (No dimension)

Since $p_L = p_w$, $\gamma_L H_L = \gamma_w H_w$ or $\gamma_w S_L H_L = \gamma S_w H_w =$ or $H_L S_L = H_w$

Hence the pumping head H_L must be converted to H_W before selecting the pump. Change of density (ρ_T) due to change in temperature can be determined by the following equation.

$$\rho_T = \rho_{15} / (1 + \beta_T (T - 15), \quad (1.11)$$

where ρ_{15} – Density at 15°C T – The pumping Temperature and β – Density coefficient.

The values of β_T are given in Table 1.4.

Table 1.4: β_T as a function of density

P_{15}	0.7	0.8	0.85	0.9	0.95
β_T	82×10^{-5}	77×10^{-5}	72×10^{-5}	64×10^{-5}	60×10^{-5}

Power of the pump $N = \gamma QH$ / (const. η). Since γ changes, power changes. However when density reduces, flow rate slightly increases, due to the increase in flow velocity in impeller and casing. Correspondingly flow rate slightly reduces when density increases.

1.5.3 Effect of Viscosity

Viscosity is explained as the resistance to flow. Viscosity is the one which produces hydraulic friction in fluids. Higher the viscosity, higher the hydraulic losses and correspondingly the power required by the pump increases. In other words, the efficiency of the pump is reduced.

Viscosity for water is considered as the standard unit and is equal to 1 centi poise (= 0.01 gm sec/cm) or 1 centi stoke (= 0.01cm²/sec). Viscosity for other liquids is always compared with respect to that of water. Determinations of hydraulic losses by theoretical means are extremely difficult. The values are determined experimentally and empirical relations are developed. Still more difficult, if the liquid is other than water. Experiments were conducted with different solutions of different viscosities and a general relationship with that of water parameters are obtained. Figure 1.20(a) gives the result of the variation of head, discharge and efficiency due to variation of viscosity.

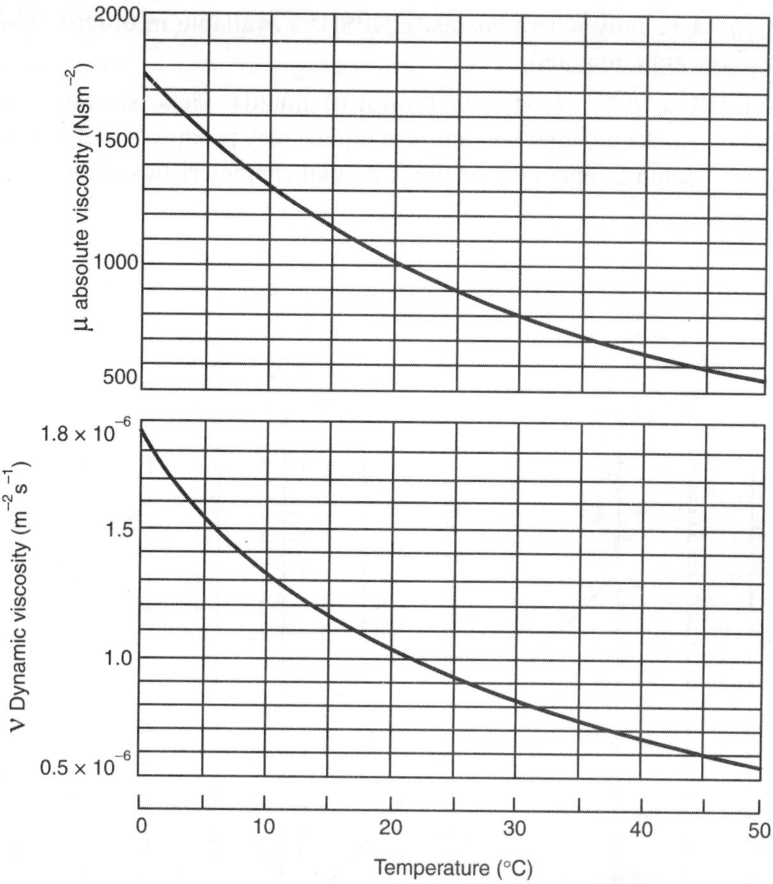

Fig. 1.18: Viscosity of water at 101 325 N/m^2

1.6 PUMP PERFORMANCE WHEN HANDLING VISCOUS LIQUIDS

1.6.1 Performance Correction Chart for Viscous Liquids and its Limitations

The Fig. 1.20(a) provides the means of determining the performance of a rotodynamic pump handling viscous liquid, when its performance on water is known. The limitations are

(*a*) Use only for pumps of conventional hydraulic design, in the normal operating range, with open or closed impellers. Do not use for mixed flow or axial flow pumps or for pumps of special hydraulic design for either viscous or non-uniform liquids.

(b) Use only where adequate NPSH is available in order to avoid cavitation; and

(c) Use only Newtonian (uniform) liquids. Gels, slurries, paper stock and other non uniform liquids may produce widely varying results, depending on the particular characteristics of the liquids.

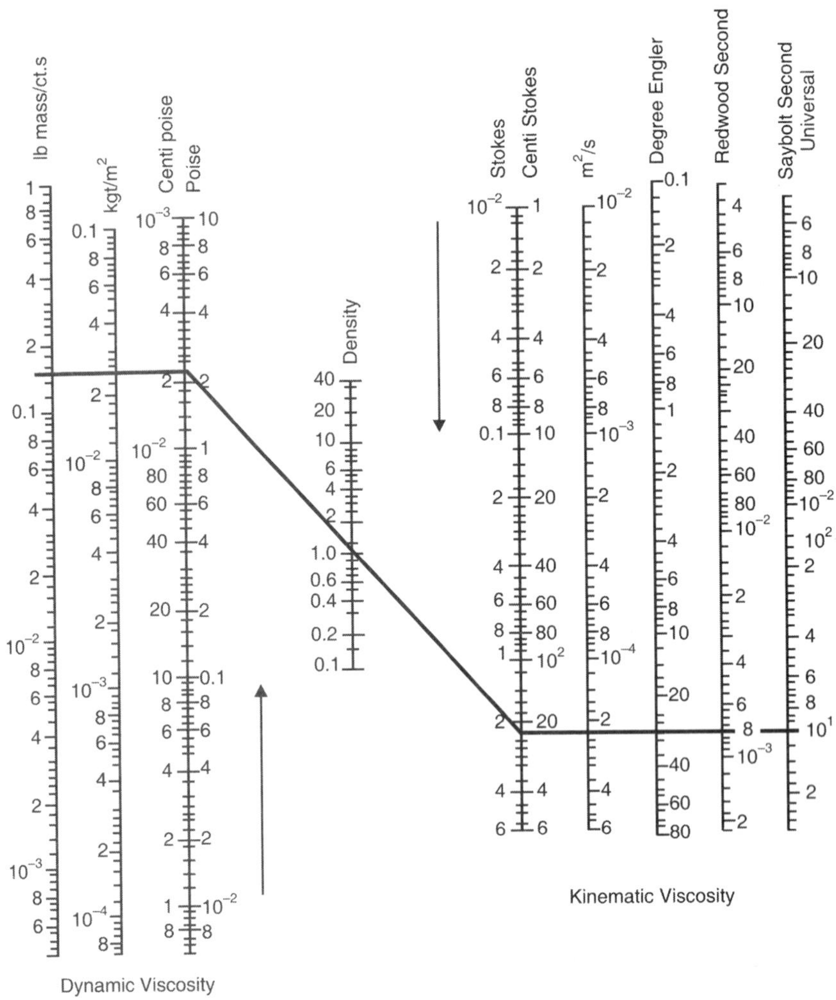

Fig. 1.19: Chart for viscosity conversion

1.6.2 Symbols and Definitions

Q_v = Viscous capacity, m^3/hr (the capacity when pumping a viscous liquid);

H_v = Viscous head, meters (the total head (H) when pumping a viscous liquid);

η_v = Viscous efficiency, percent (the efficiency when pumping a viscous liquid)

N_v = Viscous power, kW (the power required by the pump for the viscous condition)

Q_W = Water capacity, m³/hr (the capacity; when pumping water);

H_w = Water head, meters (the total head (H); when pumping water);

η_w = Water efficiency, percent (the efficiency; when pumping water);

S = Specific gravity

K_Q = Capacity correction factor;

K_H = Head correction factor;

K_η = Efficiency correction factor;

Q_{opt} = Water capacity at which maximum efficiency is obtained.

1.6.3 Equations to Determine the Viscous Performance of the Pump, for the known Water Performance

$$
\begin{aligned}
&(a)\ Q_v = K_Q \times Q_w \\
&(b)\ H_v = K_H \times H_w \qquad (1.12)\\
&(c)\ \eta_v = K_\eta \times \eta_w \\
&(d)\ P_{vis} = S_o \gamma_w Q_v H_v
\end{aligned}
$$

K_H, K_Q and K_η are determined from the Fig. 1.20, based on the water performance. The following equations are used for approximating the water performance when the desired viscous capacity and head are known and the values of K_Q and K_H, shall be estimated from Fig. 1.20. Using Q_v and H_v, corresponding water parameters are determined from the equation, Q_w (approx) = Q_v/K_Q and H_w (approx.) = H_v/K_H.

1.6.4 Selection of a Pump for a given Head, Capacity and Viscosity of the Liquid

(*a*) Given the desired capacity and head of the viscous liquid to be pumped, the viscosity and specific gravity at the pumping temperature, the chart (Fig. 1.20) may be used to find approximate equivalent capacity and head when pumping water.

(*b*) Enter the chart (Fig. 1.20) at the bottom with the desired viscous capacity (Q_v) and proceed upward to the desired viscous head (H_v) in meters of liquid. For multistage pumps, use head per stage. Proceed horizontally (either left or right) to the fluid viscosity, and then go upward to the correction curves. Find the values of K_H, K_Q, K_η for the values of water efficiency of the pump. Divide the viscous capacity (Q_v) by capacity correction factor (K_Q) to get the approximate equivalent water capacity (Q_W) (approx). Divide the viscous head (H_v) by the head correction factor (K_H) from the curve marked $1.0 \times Q_{opt}$, to get the approximate equivalent water head (H_W). Using this new equivalent water head capacity point, select a pump in the usual manner. From the viscous efficiency K_η, and the water power the corresponding viscous power may then be calculated.

(*c*) This procedure is approximate as the scales for capacity and head on the lower half of the figure are based on the water performance. However, this procedure has sufficient accuracy for most pump selection purposes.

1.6.5 Determination of Pump Performance for a Viscous Liquid, when Water Performance is known

(*a*) Given the complete performance characteristics of a pump handling water, to determine the performance when pumping a liquid with a specified viscosity.

(*b*) From the efficiency curve of the water parameters of the pump, locate the water capacity ($1.0 \times Q_{opt}$) at which maximum efficiency is obtained. From this capacity, determine the capacities ($0.6 \times Q_{opt}$), ($0.8 \times Q_{opt}$) and ($1.2 \times Q_{opt}$). Enter the chart (Fig. 1.20) at the bottom with the capacity ($1.0 \times Q_{opt}$), go upward to the head developed in one stage (H_W) at this capacity, then horizontally (either left or right) to the desired viscosity, and then proceed upward to the various correction curves. Read the values of K_H, K_Q and K_η for all four capacities. Multiply each head by its corresponding head correcting factor to obtain the corrected heads for viscous liquids. The same procedure is followed to obtain the corrected quantity for viscous liquids multiply each efficiency values by K_η to obtain the corrected efficiency values for the viscous liquids, at the corresponding corrected capacities.

(*c*) Plot corrected head and corrected efficiency against corrected capacity, drawing a smooth curve through these points. The shut-off head can be taken to be the same as that of water [Fig. 1.20(*b*)].

(*d*) Calculate the viscous power (N_v) from the formula given above.

(*e*) Plot these points and draw a smooth curve through them which should be similar to an approximately parallel to the power (N_w) curve for water. Example: Given the performance of a pump obtained by test on water, plot the performance of this pump when handling oil with a specific gravity of 0.90 and a viscosity of 210 CP at pumping temperature. On the performance curve locate the best efficiency point which determines Q_{opt}. In this case, it is 47 lit/sec. Tabulate capacity, head and efficiency for (0.6 × 750), (0.8 × 750), (1.2 × 750). Using 47 lit/sec, 30 meters head and 210 CP, enter the chart and determine the correction factors. These are tabulated in sample performance table. Multiply each value of head, capacity and efficiency by the corresponding correction factor to get the corrected values. Using the corrected values, and the specific gravity, calculate power.

Table 1.5: Viscous parameter determination from water parameters

% Q_W values / Parameters		0.6	0.8	1.0	1.2
Flow rate m³/hr	Q_W	90	120	150	180
$Q_v = K_Q Q_W$	Q_v	88.2	117.5	147	176.5
Total head m	H_W	44	42	40	36
$HV = K_H H_W$	H_v	43.2	40.8	38	33.5
Efficiency %	η_W	70	78	80	77
$\eta_V = k_\eta \eta_w$	η_v	49	54.5	56	54
Input power kW	N_W	15.7	17.9	20.9	23.1
	N_v	21.6	24.6	27.6	29.8

The following symbols and definitions shall be used:

Q_{ps} = Capacity when pumping paper stock in m^3/hr
H_{ps} = Total head when pumping paper stock in meters
η_{ps} = Efficiency (percent) when pumping paper stock
N_{ps} = Power required by the pump when handling paper stock
Q_w = Water capacity in m^3/hr (the capacity when pumping water)
H_w = Water head in meters (the total head when pumping water)
η_w = Water efficiency (percent) (the efficiency when pumping water)
K_{HC} = Head capacity correction factor
K_η = Efficiency correction factor and
S = Specific gravity.

Figures 1.20(a) and 1.22 provide the head capacity correction factor (K_{HC}) for different percent water capacity at best efficiency point. Q_w, H_w and η_w shall be known when the water performance of the pump is known.

The following equation shall be used for determining the pump performance for handling paper stock when water performance of the pump is known:

$$N_{ps} = \frac{S_{ps}\gamma_w Q_{ps} H_{ps}}{\eta_{ps}} \tag{1.13}$$

where, $Q_{ps} \times H_{ps} = Q_w \times H_w \times K_{HC}$ and $\eta_{ps} = \eta_w \times K_\eta$

1.6.6 Effect of Viscosity on Performance

The performance of water pumps, when used for viscous liquid pumping, changes considerably from water performance. Many authors already worked on this area. Prof. A.A. Burdhakoff, Prof. R.W. Sheshenko, Prof. D.A. Suhanoff, Prof. B.D. Baklanoff, Prof. M.D. Aizentein, Prof. A.T. Ippan | 4 |, Prof. A.J. Stepanpoff | 112 | are a few authors to be mentioned.

In chemical industries, centrifugal pumps, of n_s = 80 to 135, are used for pumping viscous liquids. Spiral casings are used, instead of diffusers, in most of the pumps, because, when pumping viscous liquids, the flow velocity in pump parts is less than that for water pumping. Coefficient of reaction ρ = 0.7 and number of blades Z = 5 to 6, depending upon the viscosity of the pumping liquid. Fig. 1.21 gives the test results of a pump of n_s = 85, for different viscous liquid pumping.

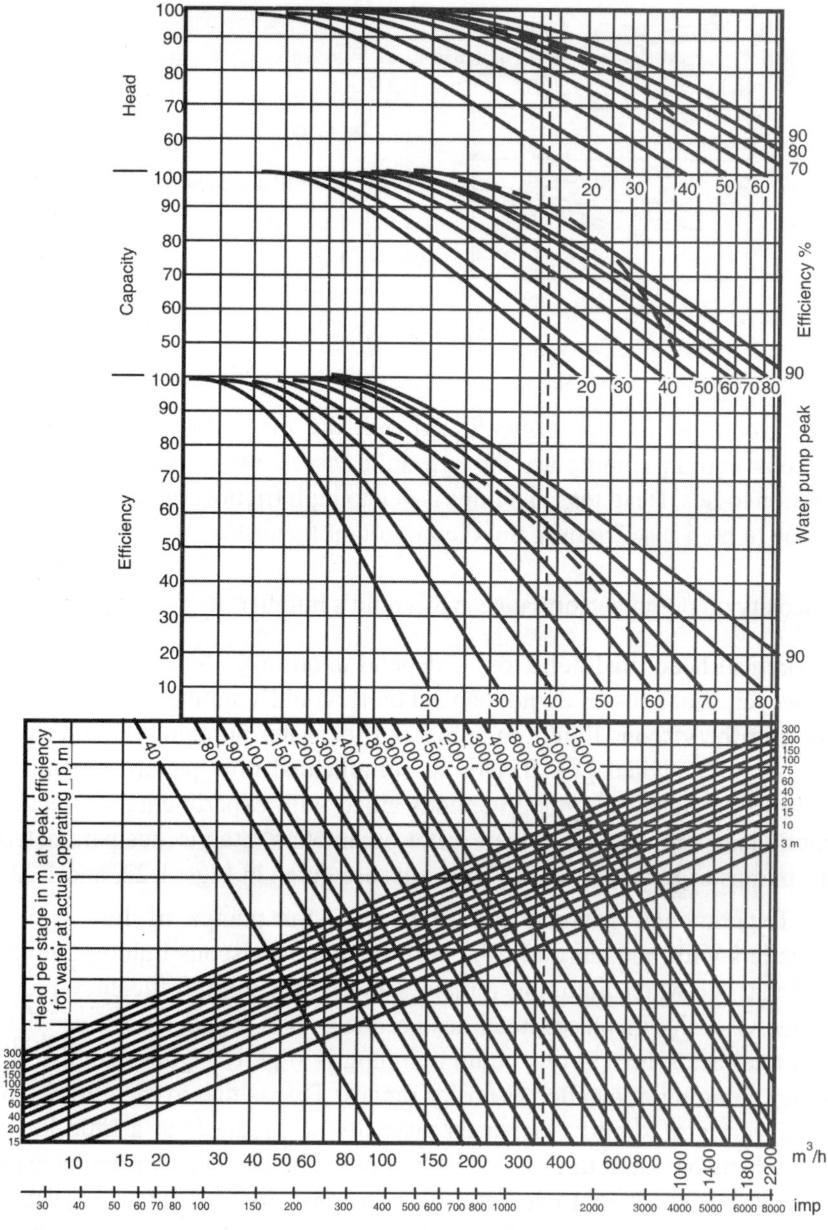

Fig. 1.20: (a) Viscosity conversion chart

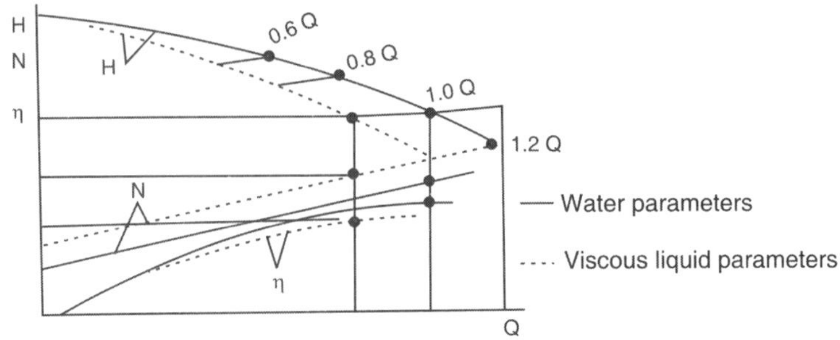

Fig. 1.20: (b)

Prof. D.A. Suhanoff mentioned that the major losses involved in viscous liquid pumping are the hydraulic losses due to friction and disc friction losses. Head loss in pumps is due to high friction, while power loss is due to high disc friction. So, the important factor to devide the effect of viscosity on pump performance is Reynold's number $R_e = \frac{\omega r^2}{\nu}$, since the theoretical head, hydraulic losses depend upon only Reynolds's number. However, shock losses do not depend on Reyonld's number. Shut off head (H_0) for the viscous liquid pumping and for water pumping remain same *i.e.*, $H_{ov} = H_o$, when $Q = 0$. Due to complicated flow pattern prevailing in flow passages, analytical methods aided with experimental correction coefficients are used to convert water parameters into viscous parameters for pumps of $n_s = 50$ to 130. The results are given in Figs. 1.22, and 1.23.

Experimental analysis indicates, that, for pumps of low n_s and impellers with both shrouds, the performance of viscous liquid pumping and that for water pumping remain same at maximum efficiency point. When viscosity increases, total head reduces. It is found, sometimes that, that total head, for viscous liquid pumping is slightly higher than that for water pumping at smaller viscosity ranges. This is due to the absence of axial vortex at slightly increased viscosity. C_{u2} reduces at a little lower rate when compared to $C_{u2\infty}$, which not only compensates for the increase in hydraulic friction losses and also increase in total head. However, this compensation cannot be made, when viscosity increases further.

The similarity laws cannot be applied directly, due to insufficient data for viscous liquid pumping, since a change in speed, changes Reynold's number. However, for low viscous liquids *i.e.*, for liquids with high

Reynold's number having ν = 5000 strokes, the performance of viscous liquid pumping coincides with that of water pumping, even when speed changes in a wide range. Model analysis must be considered including Reynold's number also.

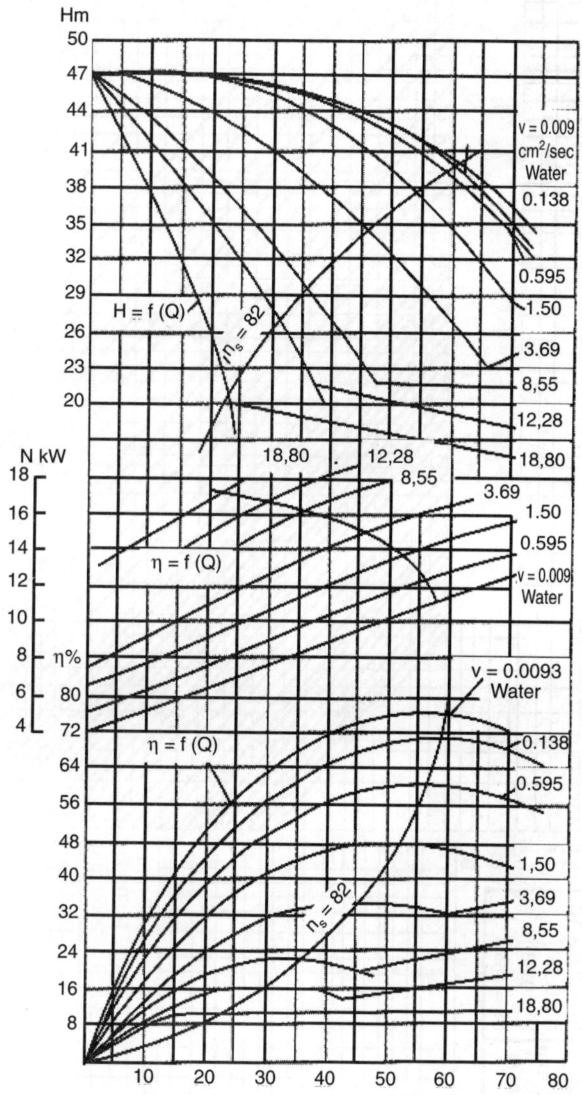

Fig. 1.21: Actual performance variation due to change viscosity of liquids test results conducted on the same pump n_s = 85 n = 2875 rpm

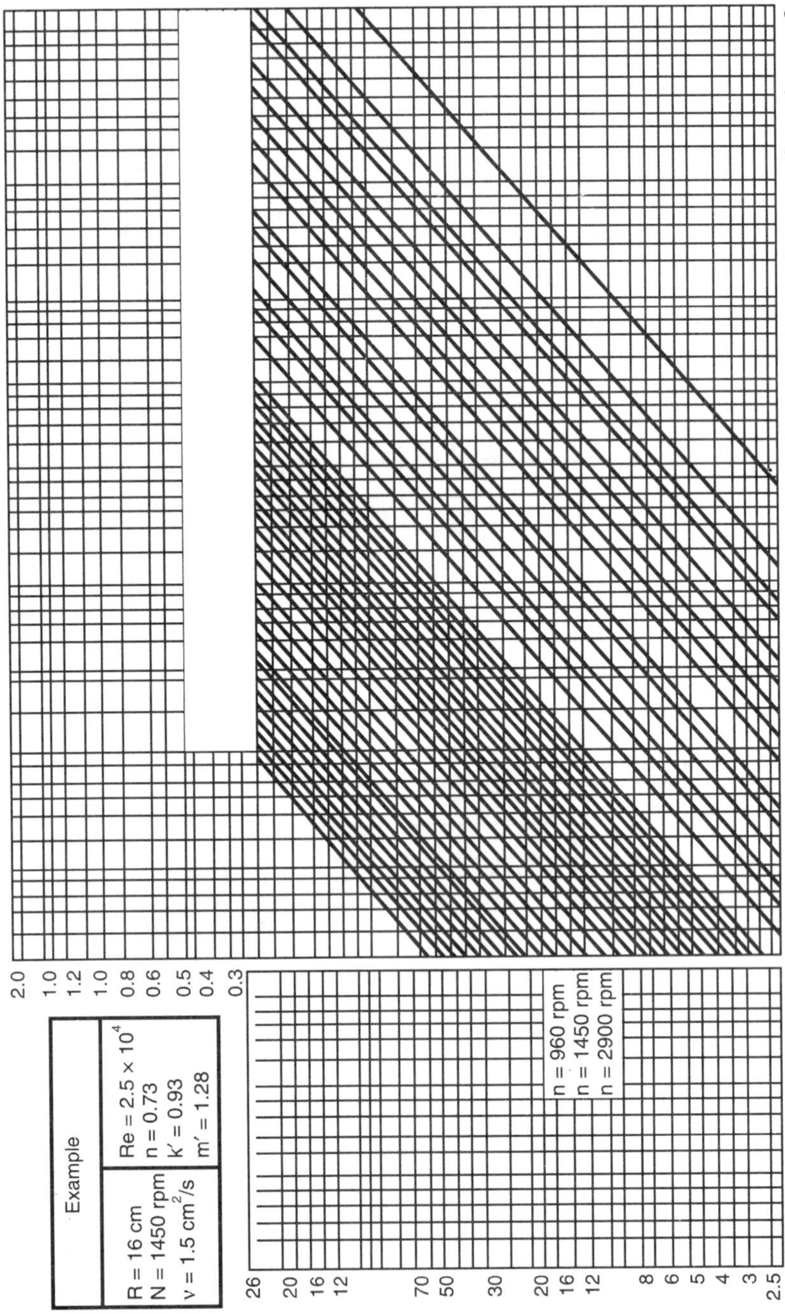

Fig. 1.22: Monogram $k', m', \eta' = f(R_e)$ to convert water parameters to equivalent liquid parameters

In order to compare the characteristics of different viscous pumping liquid, prof. D.A. Suhanoff gave a relation between head coefficient $k\ (= K_H) = \dfrac{H}{n^2 D^2}$, Power coefficient $m\ (= K_N) = \dfrac{N}{n^3 D^5}$, and efficiency (η) as a function of Reynold's number (R_e). The results, obtained are given in a monogram $k', m', \eta' = f(R_e)$ (Fig. 1.22). These results, perfectly agree with the experimental results of different viscous fluid pumping for the same pump and same viscosity for different pumps.

The variation in k', m' and $\eta' = f(R_e)$ is attributed to increase the friction losses and disc friction losses especially when $R_e < 7 \times 10^3$. At low Reynold's number η', k' and m' reduces to a greater extent than. When $R_e > 3 \times 10^5$. Calculation by dimensional analysis gave more accurate results. The error will be more, especially, when the speed is changed.

Comparing the pump characteristics for different Reynold's numbers (R_e), when unit quantity $K_Q = Q/nD^3$ is constant, Prof. Suhanoff, suggested that relative efficiency ratio (η'), relative power ratio (m'), and relative total head ratio (k') can be used for all practical purposes as a function of Reynold's number (R_e) *i.e.*, η', m', and $k' = f(R_e)$. These values are given as

$$\eta' = \frac{\eta_v}{\eta},\ m' = \frac{m_v}{m},\ k' = \frac{k_v}{k} \tag{1.14}$$

The values with suffix (v) attribute to viscous liquid and with out suffix for water. Figures 1.22 is a consolidated monogram, giving all the above mentioned details for easy calculation. This monogram has been developed based on the experimental results from many authors.

The procedure is as follows:

1. A few points, very near to the maximum efficiency point are selected on water performance graph of the pump and η, m and k are calculated.
2. Reynold's number is calculated using the formula $R_e = \dfrac{\omega r^2}{\nu}$
3. From Fig. 1.22, the relative coefficients η', m' and k' are found out for the obtained Reynold's number.
4. Using equation (1.14) the values k_v, m_v, η_v are determined.
5. The head and power of the viscous liquid can be calculated from the formula $H_v = k_v n^2 D^2$ and $N_v = m_v n^3 D^5$ and $\eta_v = \eta\eta'$.

6. Shut off head H_0' when $Q = 0$, is obtained from the pump performance graph for water. $H_{ov} = H_v'$

Another method of calculation is:

1. Under constant speed, and the change in pumping liquid viscosity, the (*H-Q*) for water pumping changes, in such a way that the specific speed for both liquid pumping remains same (Fig. 1.21).

$$n_s = \frac{3.65n\sqrt{Q}}{H^{\frac{3}{4}}} = n_{sv} = \frac{3.65n\sqrt{Q_v}}{(H_v)^{\frac{3}{4}}} \tag{1.15}$$

The (*H-Q*) curve drops down more when viscosity increases (Fig. 1.21). Suffix (ν) is for viscous liquid and without suffix is for water.

From the equation

$$\frac{Q}{Q_V} = \left(\frac{H}{H_V}\right)^{\frac{3}{2}} \tag{1.16}$$

From the equation, it is seen that the coefficients, K_H, K_N, K_η and K_c (Fig. 1.22) obtained at maximum efficiency point, remain same for all other selected points of water pump performance. All other values can be obtained from the above correction coefficient.

2. Shut off head (H_0) for water pumping is taken from the water pump performance graph. This is same for viscous fluid pumping also, since shut off head is independent of viscosity of pumping liquid.

Figure 1.21, it is seen that under viscous liquid pumping, the power required increases more or less by the same value for a wide range of flow rate.

For water pumping $N = \dfrac{\gamma QH}{\text{Constant } \eta_v}$

For viscous liquid pumping, $N_V = \dfrac{\gamma_v Q_v H_v}{\text{Constant } \eta_v}$

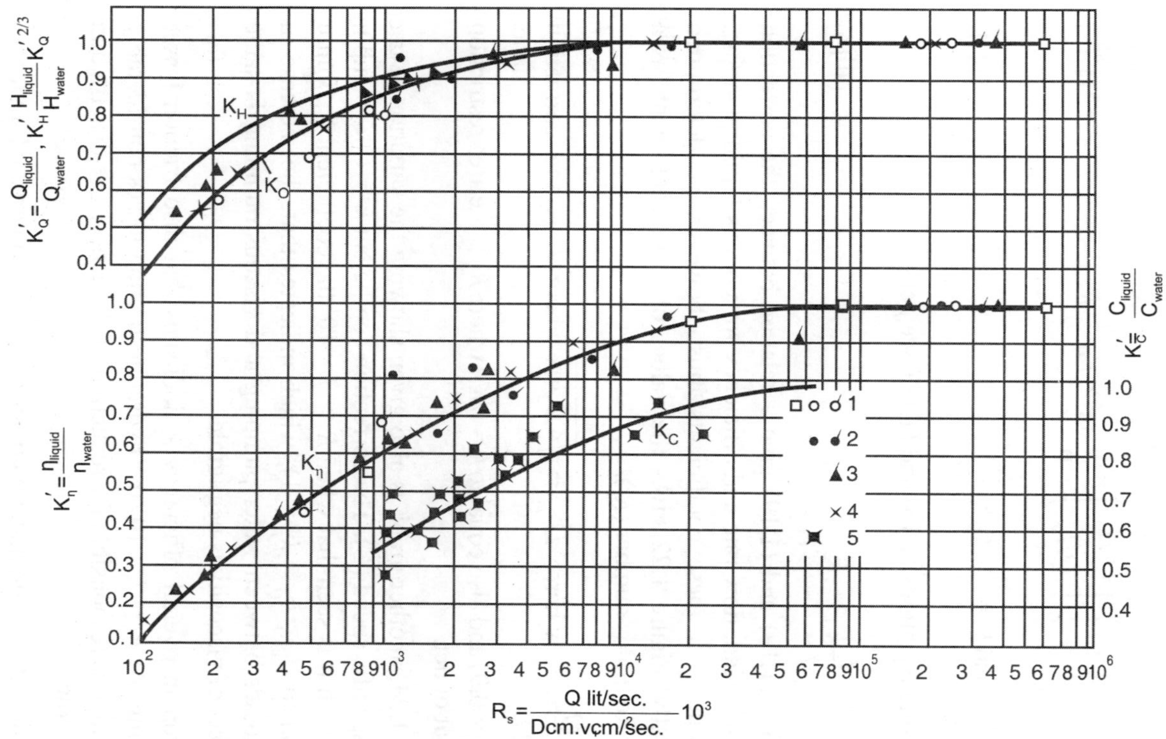

1—as per Ajenshtein (n_{si} = 51,60 to 70); 2—as per Suhanoff (n_{si} = 82 to 130); 3—as per Ippen (n_{si} 90 to 115); 4—as per Stepanoff (n_{si} = 82); 5—as per Ajenshtein for cavitation (n_{si} = 60 to 100).

Fig. 1.23: Coefficients K'_Q, K'_H, K'_η, and $K'_C = f(R_e)$

Combining these two, $N < N_V$

we get $$\frac{QH}{\eta} < \frac{\gamma_v}{\gamma} \times \left(\frac{Q_v H_v}{\eta_v}\right) \tag{1.17}$$

$$\left(\frac{\gamma_v}{\gamma}\right) \times \left(\frac{Q_v H_v}{QH_v}\right) > \frac{\eta_v}{\eta} \text{ or } k_Q k_H \frac{\gamma_v}{\gamma} k_\eta \tag{1.18}$$

where $k_Q = \dfrac{Q_V}{Q}$ and $k_\eta = \dfrac{\eta_v}{\nu}$ $k_H = \dfrac{H_V}{H}$,

Combining equations (1.16) and (1.18), we get

$$\frac{\gamma_v}{\gamma} k_H^{\frac{5}{2}} > k_\eta \tag{1.19}$$

Equations (1.18) and (1.19) give a relation between the correction coefficient for head, flow rate and efficiency for viscous liquid pumping. Absolute values of these coefficients must, however, be determined experimentally. Figure 1.22 gives the values of these coefficients as a function of Reynolds number (R_e).

$R_e = \dfrac{Q_{nor}}{D_{equ} V}$, where $D_{equ} = \sqrt{4 D_2 b_2 k}$ where k is the correction coefficient (≈ 0.9) and also equal to $k = 1/K_2$, where K_2 is the area restriction coefficient at outlet.

In Fig. 1.23, another graph is also given showing the variation of the cavitation specific speed (C). The correction coefficient $k_c = C_v/C = f(R_e)$. From the graph it is seen that when $R_e > 7 \times 10^3$ the coefficient k_Q, k_H are nearly equal to (*i.e.*, $k_Q = k_H = 1$), which indicates that, the increase in hydraulic losses between water pumping and viscous pumping is same *i.e.*, the effect of viscosity is negligible.

The drop in overall efficiency k_η indicates that disc friction losses are increasing. At $R_e < 7 \times 10^3$, the hydraulic friction losses considerably increase, which is responsible for reduction in efficiency. The disc friction loss will be low.

In practice, it is found that, pump performance comparison for viscous liquid pumping with that of water with respect to Reynolds's number, gives an error of ± 5% for $n_s = 85$ to 130.

1.6.7 VISCOSITY BLENDING CHART

The standard A.S.T.M. chart Fig. 1.24 provides a rapid and accurate method of finding the viscosity of a blend of two liquids; or the proportions needed to produce a blend of required viscosity. Plot the viscosity of the low viscosity component on the left hand vertical scale (kinematic viscosity *A*) and the viscosity of higher viscosity component on the right hand scale (kinematic viscosity *B*). Join by a straight line. Project upwards or downwards from the corresponding proportion of components *A* or component *B*, respectively. Where this cuts the diagonal, project across to left hand vertical scale to read viscosity of the blend. Alternatively, to find the proportions for required blend viscosity, project across from the left hand scale at the blend viscosity value required to cut the diagonal. Project up (or down) to read the proportions of the components.

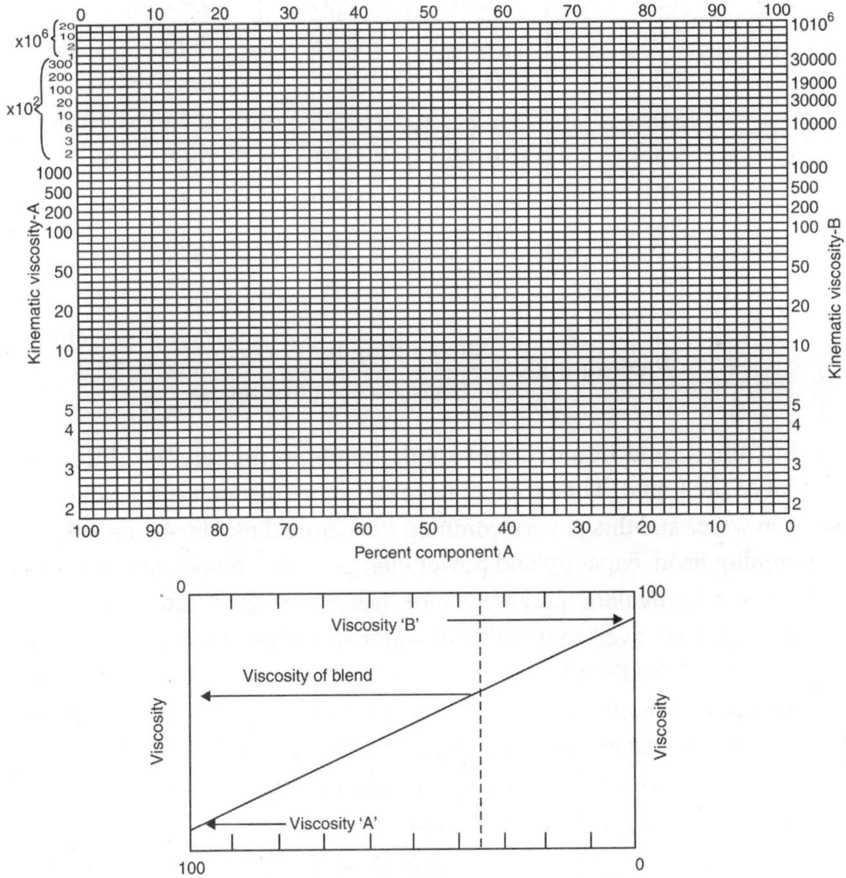

Fig. 1.24: Viscosity blending chart

1.7 CONSISTENCY

In chemical industries pumps handle liquids with solids in suspension, which is called "Two Phase Flow". Apparently the average specific gravity

Table. 1.6: Conversion factors for API/scan pumps/centrifugal pulp pumps, based on data for water and stock at different concentrations (consistency)

Pulp Consistancy %	E_Q	E_H	E_Z
1	0.99	1.00	0.99
2	0.99	1.00	0.99
2.5	0.98	1.00	0.98
3.0	0.97	1.00	0.97
3.5	0.96	0.99	0.95
4.0	0.92	0.98	0.90
4.5	0.87	0.97	0.85
5.0	0.8	0.95	0.76
5.5	0.72	0.93	0.67
6.0	0.62	0.90	0.56
6.5	0.52	0.87	0.45
7.0	0.42	0.83	0.35

$$\text{Water rating} = \frac{\text{Stock rating for } Q \text{ or for } H}{\text{Constant for } Q \text{ or for } H}$$

of the mixture (liquid-liquid, liquid-solid, liquid-gas mixtures) is different from that of the liquid. The problem becomes more difficult if the liquid is other than water and this is very common in chemical plants. As a result of this, pumping head, capacity and power change. Since pumps are selected based on the equivalent parameters of water, the parameters of such mixtures are also converted into that of water with the help of a conversion factor called "CONSISTENCY". The solid content of mixture expressed in percentage by volume or weight (or specific gravity) of mixture is often referred to as consistency (or concentration). It is defined as that property of a material by which it resists permanent change of shapes and it is defined by the complete flow force relations.

1.8 HYDROSTATIC HEAD OF LIQUIDS, OTHER THAN WATER

In designing the equipments and piping systems it is often necessary to convert feet of head of a liquid into psi. The relationship depends on the specific gravity of the liquid. This monogram quickly relates these three factors and lists the specific gravity for a large variety of industrial liquids.

Example: Find the pressure under 9 ft of ethylene. First, find the specific gravity of ethylene on the left hand scale; connect it with a straight line to 9.0 on the right hand scale. Find the answer at the intersection with the center scale, 3.9 psi.

Here is how to calculate loss of head due to friction. First find the diameter and the discharge of water you require. Then find from Table 1.6, what will be the loss of head for the pipe diameter and rated output. This will be for a pipe line of 100 feet long. Calculate the length of your pipe laid in the field. Then work out the frictional loss for your pipe line length.

Loss of head in fittings. There is one more source of loss of head. And that is due to the fittings such as bends, foot valve, 'T' piece or sharp elbow, check valve, strainer etc. All these are obstructions to the free flow of water, and therefore they entail loss of head.

For calculating the loss of head of each kind of fitting we convert the loss in terms of equivalent pipe length. In Table 1.9 is given the pipe length equivalent for each kind of fitting and for the various diameters.

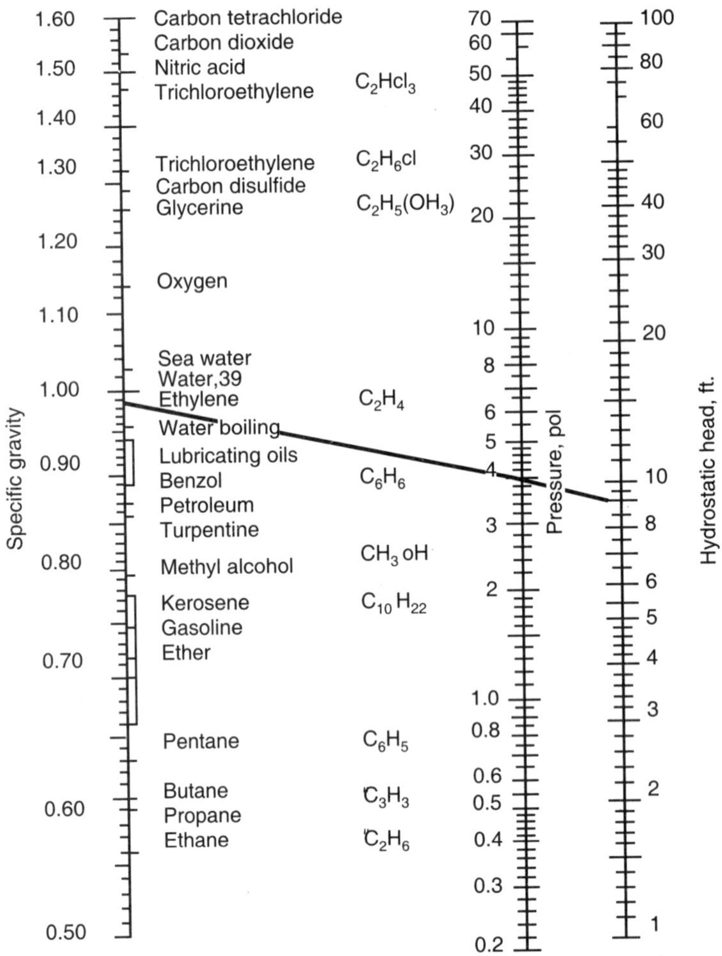

Fig. 1.25: Hydrostatic head of liquids

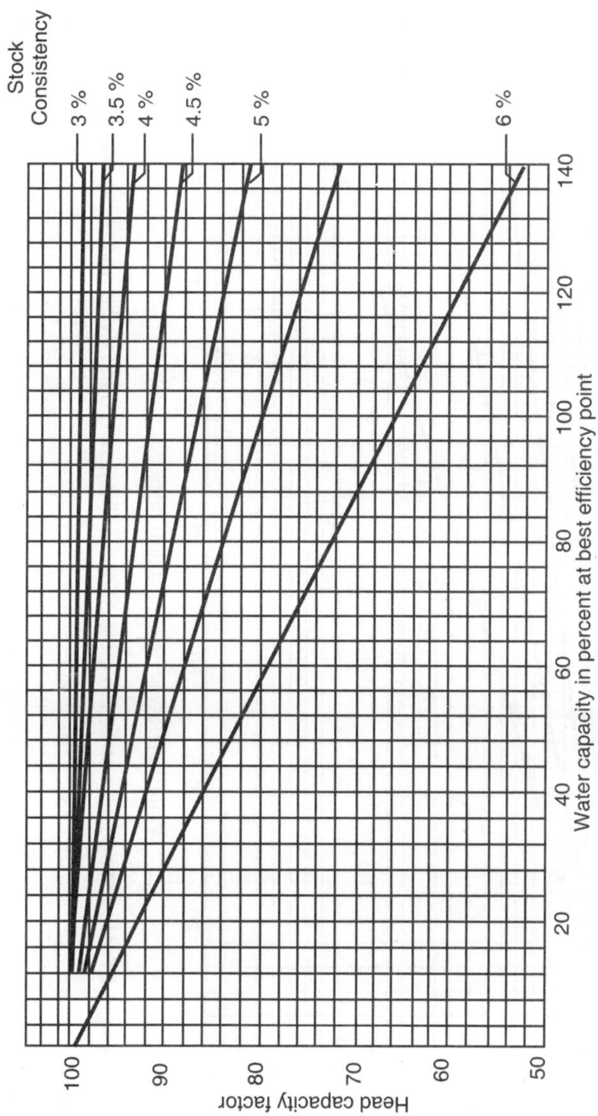

Fig. 1.26: Head capacity correction factors for chemical stock (Soda, sulphite, sulphate, draft, semi-chemical)

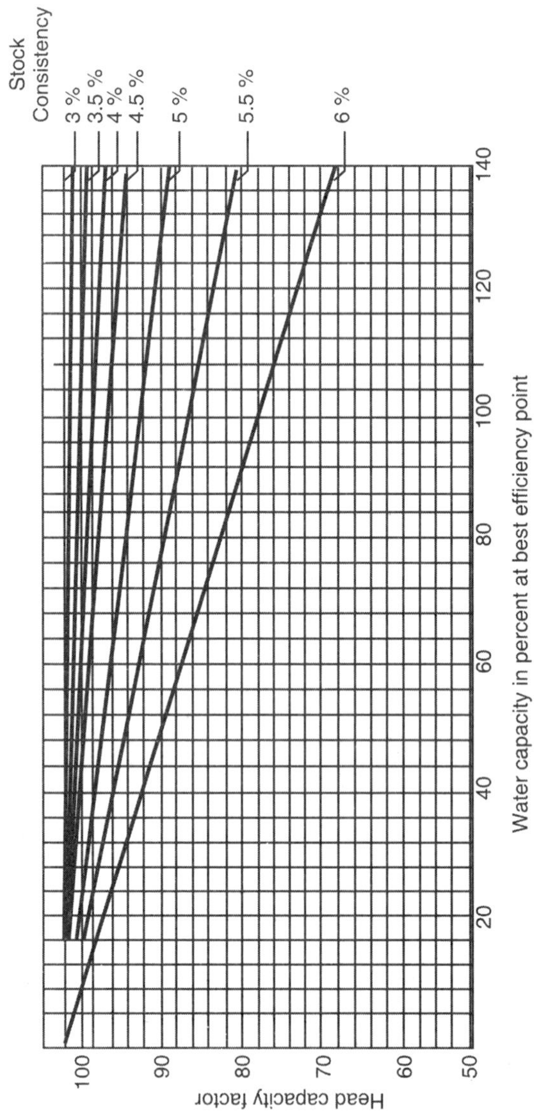

Fig. 1.27: Head capacity correction factors for mechanical ground wood and reclaimed stock

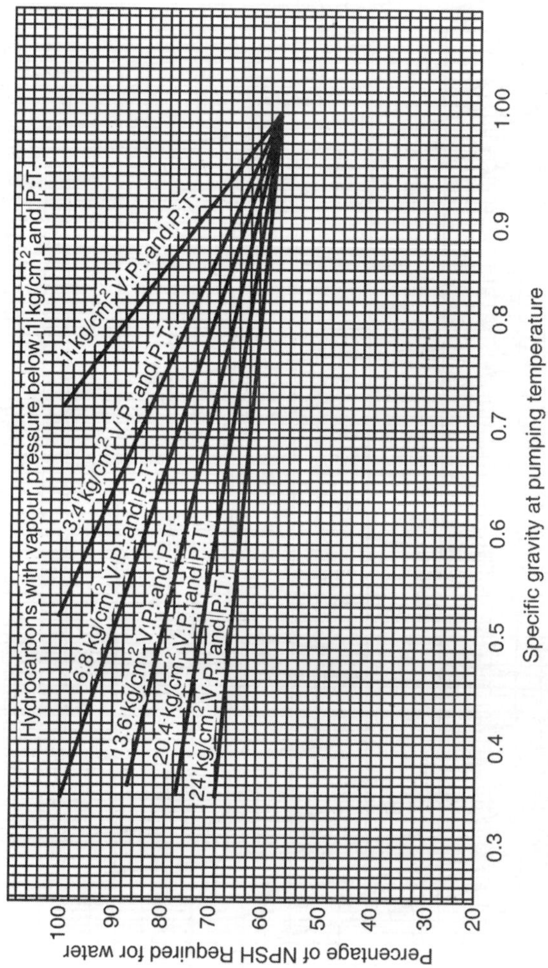

Fig. 1.28: NPSH correction chart for hydrocarbons (Not to be used for other liquids)

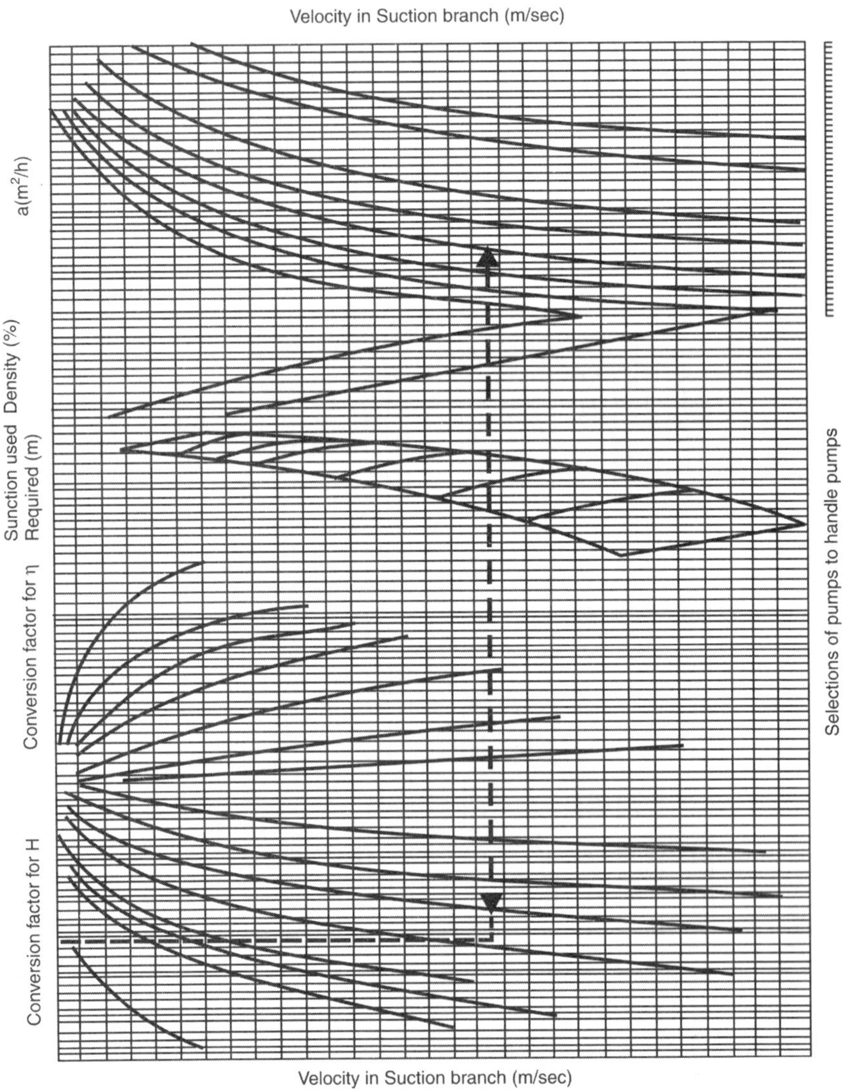

Fig. 1.29: Selection of pump to handle pulps

Table. 1.7: Friction loss of head in feet for each 100' length of pipe

Discharge in Imp. Gal/ min	Pipe size in inches												
	¾	**1**	**1¼**	**1½**	**2**	**2½**	**3**	**3½**	**4**	**5**	**6**	**7**	**8**
10	50	16	4.3	1.7	0.4	0.13							
20		40	10.0	6.5	1.6	0.5							
30		75	20.0	14.5	3.5	1.1	0.47						
40			37.0	26.0	6.0	2.0	0.8						
50				40.0	9.4	3.0	1.25	0.55					
60				57.0	13.5	4.3	1.80	0.77	0.4				
70				78.0	18.0	5.8	2.4	1.05	0.55	0.17			
80					24.0	7.5	3.1	1.4	0.72	0.22			
90					30.0	9.4	3.9	1.7	0.9	0.28			
100					37.0	12.0	4.9	2.15	1.12	0.35	0.16		
110					43.0	14.0	5.7	2.55	1.3	0.38	0.19		
120					51.0	16.5	6.8	3.0	1.58	0.5	0.22		
125					56	17.8	7.4	3.2	1.7	0.53	0.24		
140					69	22.0	9.1	4.1	2.12	0.67	0.3		
150					80	26.0	10.5	4.7	2.45	0.76	0.34	0.15	
165						29.0	12.0	5.3	2.75	0.87	0.39	0.17	
175						34.0	14.0	6.2	3.25	1.00	0.47	0.2	

180	36.0	14.9	6.6	3.45	1.09	0.49	0.21	
200	45.0	18.3	8.2	4.25	1.33	0.6	0.26	
220	54.0	22.0	10.0	5.20	1.6	0.71	0.31	
225	57.0	27.0	10.1	5.3	1.7	0.75	0.32	
250	70.0	27.5	12.5	6.45	2.05	0.92	0.4	
275		34.0	15.0	7.9	2.5	1.1	0.48	
300		40.0	18.0	9.3	3.0	1.3	0.57	
325		46.0	22.5	10.1	3.5	1.5	0.66	
350		53.0	24.2	12.5	4.0	1.75	0.75	
400				16.3	5.2	2.27	0.99	0.5
450				20.0	6.5	2.8	1.2	0.62
475				21.4	7.4	3.1	1.33	0.68
500				25.0	8.0	3.4	1.5	0.76
600				34.0	11.3	4.8	2.1	1.07
625					12.4	5.2	2.13	1.15
700					15.4	6.5	2.8	1.44

Table 1.8: Friction loss of head in meters for each 30.5 m length of pipe

Discharge in lit/sec	Pipe size in inches												
	¾	1	1 ¼	1	2	2 ½	3	3 ½	4	5	6	7	8
0.76	15.24	4.88	1.31	0.52	0.12	0.04							
1.52		12.2	3.05	2.0	0.49	0.15							
2.27		22.4	6.1	4.62	1.07	0.34	0.14						
3.03			11.28	9.93	1.63	0.61	0.24						
3.79				18.19	2.87	0.92	0.38	0.17					
4.55				17.35	4.12	1.31	0.55	0.24	0.12				
5.30				23.77	5.49	1.77	0.73	0.32	0.17	0.05			
6.06					7.3	2.29	0.95	0.43	0.22	0.07			
6.82					9.14	2.87	1.19	0.52	0.27	0.09			
7.58					11.28	3.66	1.49	0.68	0.34	0.11	0.05		
8.34					13.11	4.27	1.74	0.77	0.4	0.12	0.06		
9.09					15.5	5.03	2.07	0.91	0.48	0.15	0.07		
9.47					17.07	5.43	2.26	0.98	0.52	0.16	0.075		
10.61					21.03	6.71	2.77	1.25	0.65	0.20	0.09		
11.37					24.38	7.90	3.2	1.43	0.75	0.23	0.10	0.05	
12.5						8.84	3.66	1.62	0.84	0.26	0.12	0.05	
13.26						10.36	4.27	1.89	0.99	0.31	0.14	0.06	
13.64						10.97	4.54	2.01	1.05	0.33	0.15	0.064	

15.15						13.72	5.52	2.5	1.3	0.41	0.18	0.08	
16.67						16.46	6.71	3.1	1.59	0.49	0.22	0.10	
17.05						17.37	8.23	3.2	1.62	0.52	0.23	0.10	
18.94						21.34	8.38	3.81	1.97	0.67	0.28	0.12	
20.84							10.26	4.57	2.41	0.76	0.34	0.15	
22.73							12.19	5.69	2.84	0.91	0.4	0.17	
24.63							14.02	6.56	3.08	1.07	0.46	0.2	
26.52							16.16	7.38	3.81	1.22	0.57	0.23	
30.31									4.97	1.59	0.69	0.3	0.15
34.10									6.1	1.98	0.85	0.37	0.19
36.00									6.52	2.26	0.95	0.41	0.21
37.88									7.62	2.44	1.04	0.46	0.23
45.46									10.32	3.44	1.46	0.64	0.33
47.36										3.76	1.59	0.65	0.35
53.04										4.69	1.98	0.85	0.44

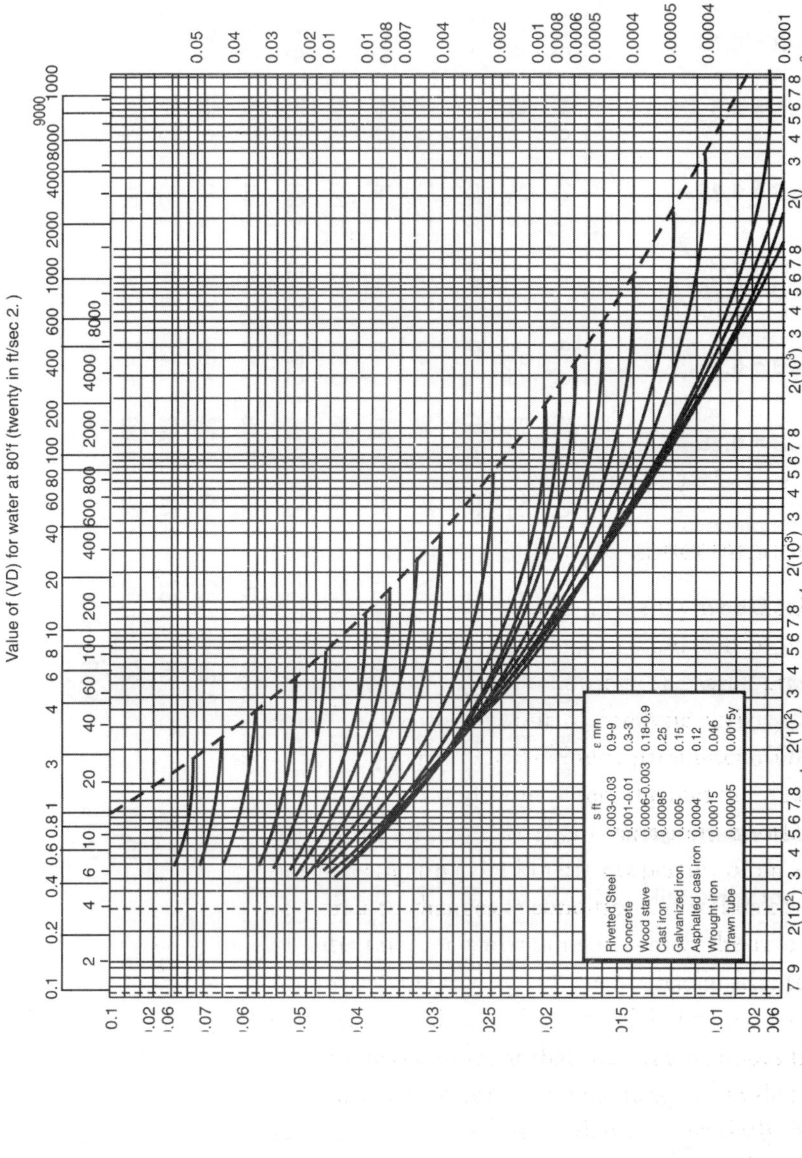

Fig 1.30: Stanton diagram - Chart to determine the frictional coefficient '*f*'

Table 1.9: Minor frictional losses due to fitting expressed as equivalent length of pipe in meters

Diameter (mm)	Ball valve fully open	Valve fully open	Foot valve fully open	Non return valve	Sieve	90° Elbow	45° Elbow	Tee piece	A/B = 0.25	A/B = 0.5	A/B = 0.75	A/B = 0.25	A/B = 0.5	A/B = 0.75	Entrance into pipe	Exit from pipe
15	4.88	0.11	1.22	1.16	3.05	0.55	0.24	1.04	0.23	0.18	0.12	0.49	0.34	0.13	0.46	0.27
20	6.40	0.14	1.52	1.59	3.66	0.69	0.31	1.37	0.31	0.24	0.14	0.64	0.43	0.14	0.61	0.37
25	8.23	0.18	1.83	1.58	4.27	0.84	0.41	1.77	0.40	0.29	0.18	0.82	0.52	0.18	0.76	0.46
32	11.28	0.24	2.13	2.74	4.88	1.14	0.52	1.79	0.49	0.40	0.24	1.10	0.70	0.24	1.07	0.61
40	13.72	0.29	2.44	3.35	5.49	1.36	0.61	2.74	0.61	0.46	0.29	1.34	0.85	0.29	1.22	0.73
50	16.76	0.38	2.74	4.27	6.10	1.62	0.76	3.66	0.76	0.55	0.37	1.65	1.07	0.37	1.52	0.91
65	19.81	0.43	3.05	5.18	6.71	1.98	0.91	4.27	0.91	0.70	0.43	1.98	1.25	0.43	1.83	1.10
80	25.91	0.53	3.66	5.79	7.62	2.50	1.16	4.88	1.13	0.85	0.49	2.44	1.55	0.49	2.29	1.37
90	28.96	0.61		6.71		2.99	1.37	5.49	1.37	1.00	0.61	2.90	1.83	0.61	2.74	1.59
100	36.58	0.73		7.67		3.66	1.52	6.71	1.52	1.13	0.70	3.35	2.13	0.70	3.05	1.83
125	44.20	0.88		10.06		4.27	1.98	8.23	1.83	1.43	0.85	4.27	2.74	0.85	3.96	2.23
150	51.82	1.07		12.19		4.88	2.29	10.06	2.29	1.74	1.07	4.88	3.35	1.07	4.57	2.74

Table 1.10: Loss in pipe fittings equivalent length in meters for various pipe fittings

Size in Inches	Bend	Foot Valve	Tee-Joint elbow	Check valve Gate valve	Strainer
¾	0.46	0.46	0.41	0.31	
1	0.61	0.67	0.82	0.31	1.22
1¼	0.76	0.76	1.07	0.36	1.22
1½	1.13	0.91	1.53	0.49	1.37
2	1.53	1.22	2.13	0.67	1.53
2½	1.95	1.53	2.74	0.85	1.83
3	2.44	1.83	3.1	1.07	1.83
3½	2.82	2.13	3.66	1.25	2.13
4	3.26	2.44	3.96	1.43	2.44
5	4.27	3.0	5.18	1.86	3.1
6	5.27	3.66	6.4	2.32	3.35
7	6.4	4.27	7.62	2.8	3.66
8	7.62	4.88	9.14	3.35	4.27

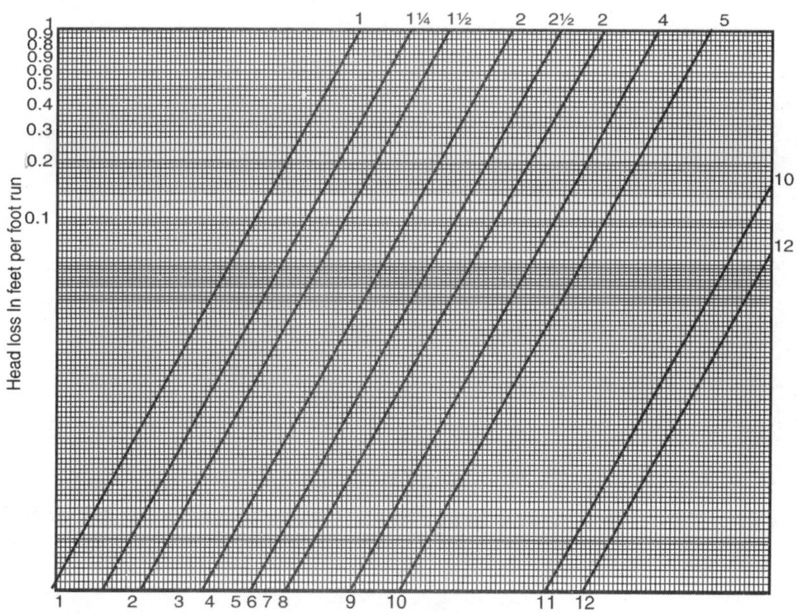

Fig. 1.31: Pipe friction loss chart

Table 1.11: Losses in pipe fittings expressed as equivalent length in feet for various pipe fittings.

Size in inches	Bend	Foot valve	Tee-piece sharp elbow	Check valve	Strainer
¾	1.5	1.5	2.0	1.0	
1	2.0	2.0	2.7	1.0	4.0
1¼	2.5	2.5	3.5	1.2	4.0
1½	3.7	3.0	5.0	1.6	4.5
2	5.0	4.0	7.0	2.2	5.0
2½	6.4	5.0	9.0	2.8	6.0
3	8.0	6.0	10.0	3.5	6.0
3½	9.3	7.0	12.0	4.1	7.0
4	10.7	8.0	13.0	4.7	8.0
5	14.0	10.0	17.0	6.1	10.0
6	17.3	12.0	21.0	7.6	11.0
7	21.0	14.0	25.0	9.2	12.0
8	25.0	16.0	30.0	11.0	14.0

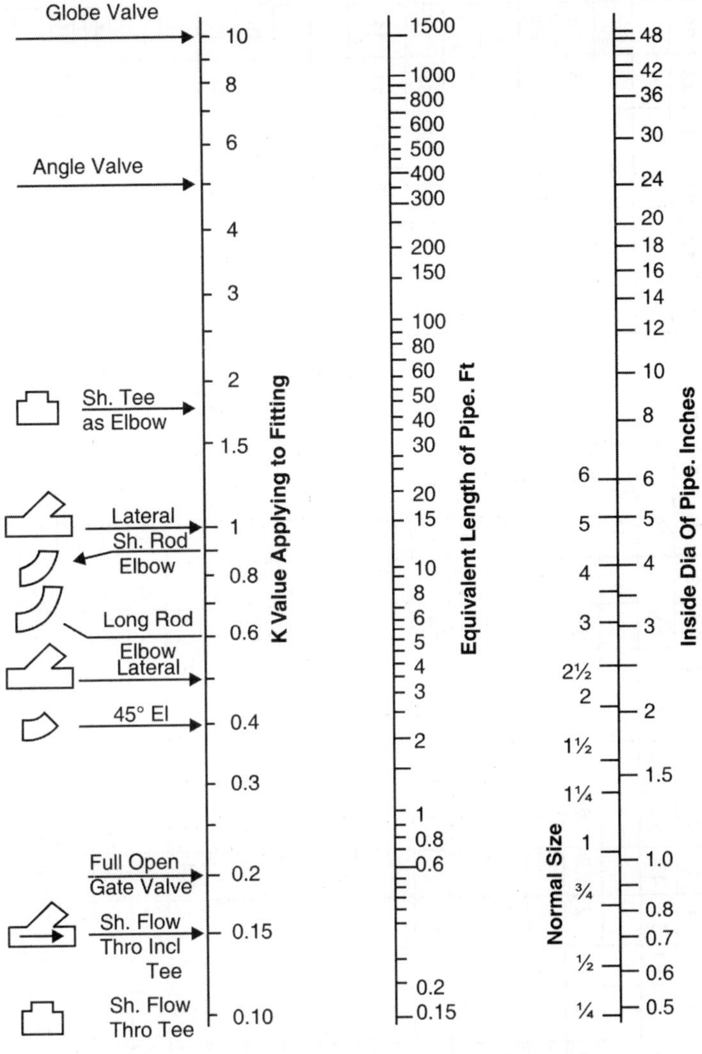

Fig. 1.32: Friction losses in fittings expressed as equivalent length of straight pipe

For pipes 10 years old, multiply the above values by 1.2

For pipes 15 to 20 years old, multiply the above values by 1.4

Table 1.12: Equivalent length of new straight pipe for valves and fittings (for turbulent flow only)

Fittings			Pipe size																			
			¼	⅜	½	¾	1	1¼	1½	2	2½	3	4	5	6	8	10	12	14	16	18	20
Regular 90° E||	Screwed	Steel	2.3	3.1	3.6	4.4	8.2	6.6	7.4	8.5	8.8	11	18									
		C.I.										9.0	11									
	Flanged	Steel			.12	1.2	1.6	2.1	2.4	8.1	8.6	4.4	6.9	7.3	6.6	12	14	17	18	21	23	25
		C.I.										3.6	4.8		7.2	2.5	12	15	17	19	22	24
Long Red. 90° E||	Screwed	Steel	1.0	2.0	2.2	2.5	2.7	3.2	3.4	3.6	3.8	4.0	4.6									
		C.I.										3.3	3.7									
	Flanged	Steel			1.1	1.8	1.10	2.0	2.3	2.7	2.8	3.4	4.2	6.0	6.7	7.0	8.0	9.0	9.4	10	11	12
		C.I.										2.8	3.4		4.7	5.7	6.8	7.8	9.6	9.8	11	11
Regular 45° E||	Screwed	Steel	.34	.52	.71	.02	1.3	1.7	2.1	2.7	3.2	4.0	5.5									
		C.I.										3.3	4.5									
	Flanged	Steel			.48	.58	.81	1.1	1.8	1.7	2.0	2.6	3.5	4.5	5.6	7.7	9.0	11	13	15	16	18
		C.I.										2.1	2.8		4.5	6.5	8.1	9.7	12	13	15	17
Tee–line flow	Screwed	Steel	.70	1.2	1.7	2.4	3.2	4.6	5.6	7.7	9.8	12	17									
		C.I.										8.8	14									
	Flanged	Steel			.68	.82	1.0	1.3	1.5	1.8	1.9	2.2	2.6	3.3	3.8	4.7	5.2	6.0	6.4	7.2	7.6	8.2
		C.I.										1.9	2.2		3.1	3.9	4.6	5.2	5.8	6.5	7.2	7.7
Tee–branch flow	Screwed	Steel	2.4	5.6	4.2	5.8	6.6	8.7	9.9	12	13	17	21									
		C.I.										14	17									
	Flanged	Steel			2.0	2.4	3.3	4.4	5.2	6.6	7.5	9.4	12	15	18	24	30	34	57	43	47	52
		C.I.										7.7	10		15	20	25	30	35	39	44	49

Fitting	Type	Material																			
180° return band	Screwed	Steel	2.3	3.1	3.6	4.4	5.2	6.6	7.4	8.5	9.3	11	12								
		C.I.										0.0	11								
	Reg. flanged	Steel			.92	1.2	1.8	2.1	2.4	3.1	3.8	4.4	5.5	7.3	8.8	12	14	17	18	21	23	25
		C.I.										3.6	4.5		7.2	8.8	12	15	17	19	22	24
	Long rad flanged	Steel			11	1.8	1.9	2.0	2.3	2.7	2.9	3.4	4.2	5.0	5.7	7.0	8.0	9.0	9.4	10	11	12
		C.I.										2.5	3.4		4.7	5.7	6.8	7.5	8.6	9.6	11	11
Globe valve	Screwed	Steel	21	22	22	24	29	37	42	54	62	70	110								
		C.I.										65	86								
	Flanged	Steel			38	40	45	54	59	70	77	94	120	160	180	210	310	300			
		C.I.										77	98		150	210	270	350			
Gate valve	Screwed	Steel	.32	.40	.56	.67	.84	1.1	1.2	1.5	1.7	1.8	2.5								
		C.I.										1.4	2.0								
	Flanged	Steel								2.6	2.7	2.8	2.9	3.1	3.2	3.2	5.2	5.2	5.2	3.2	3.2	3.2
		C.I.										2.3	2.4		2.6	2.7	2.8	2.9	2.9	3.0	3.0	3.0
Angle valve	Screwed	Steel	12.8	15	15	15	17	18	18	18	18	18	18								
		C.I.										15	15								
	Flanged	Steel			15	15	17	18	18	21	22	28	30	50	53	80	120	140	160	170	210	240
		C.I.										23	31		52	74	88	120	150	170	200	230
Swing check valve	Screwed	Steel	7.2	7.5	8.0	8.8	11	13	15	18	22	27	36								
		C.I.										22	31								
	Flanged	Steel			3.8	5.5	7.2	10	12	17	21	27	30	50	65	90	120	140			
		C.I.										22	31		52	74	80	120			

Coupling or union	Screwed	Steel	.14	.16	.21	.24	.28	.36	.38	.40	.47	.63	.85									
		C.I.										.44	.52									
	Bell mouth inlet	Steel	.04	.07	.10	.13	.18	.26	.31	.45	.52	.67	.95	1.5	1.8	2.3	2.9	3.0	4.0	4.7	5.3	6.1
		C.I.										.55	.77		1.3	1.6	2.4	5.0	5.6	4.3	5.0	5.7
	Square mouth inlet	Steel	.44	.00	.96	1.3	1.6	2.4	3.1	4.3	5.2	6.	9.6	13	16	23	20	30	40	47	55	61
		C.I.										5.5	7.7		13	18	24	30	35	45	50	57
	Re-effiesat pipe	Steel	.50	1.4	1.9	2.6	3.0	5.1	5.2	5.5	10	13	15	20	32	45	52	70	80	95	110	120
		C.I.										11	15		26	37	49	01	73	94	100	110
	Sudden En	Steel	$N = \frac{(V_1 - V_2)^2}{2}$ feet of liquid: if $V_2 = 0$ n = $\frac{V_1^2}{2}$ feet of liqiud																			
		C.I.																				

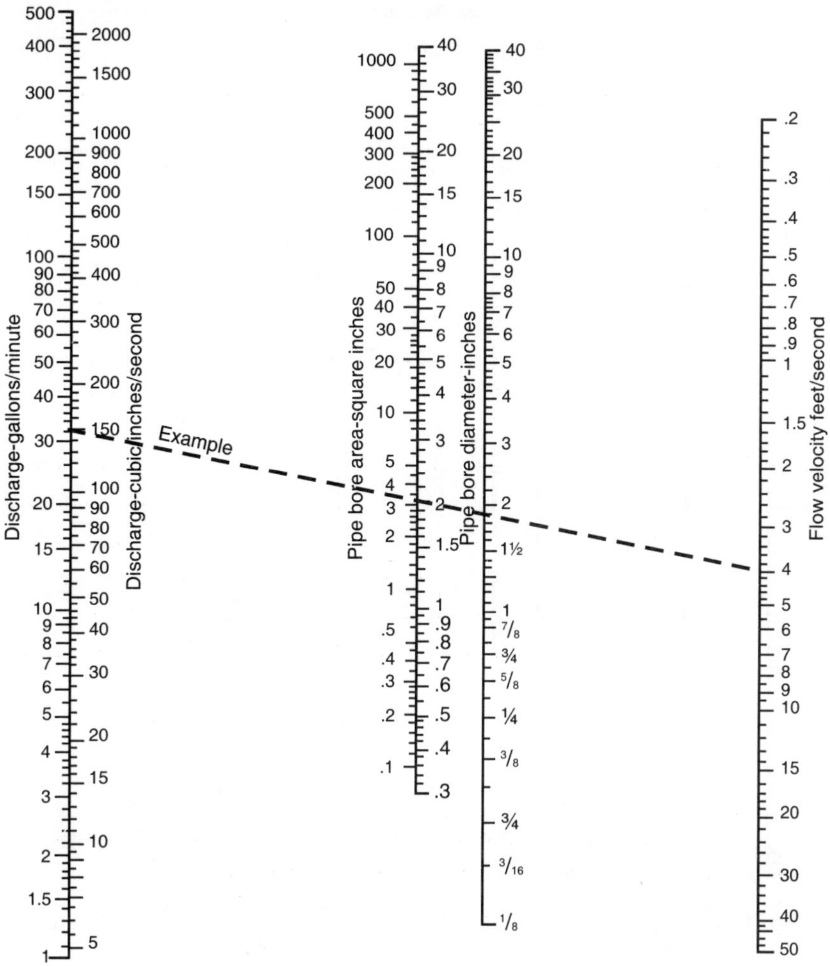

Fig. 1.33: Flow velocity discharge – Pipe size

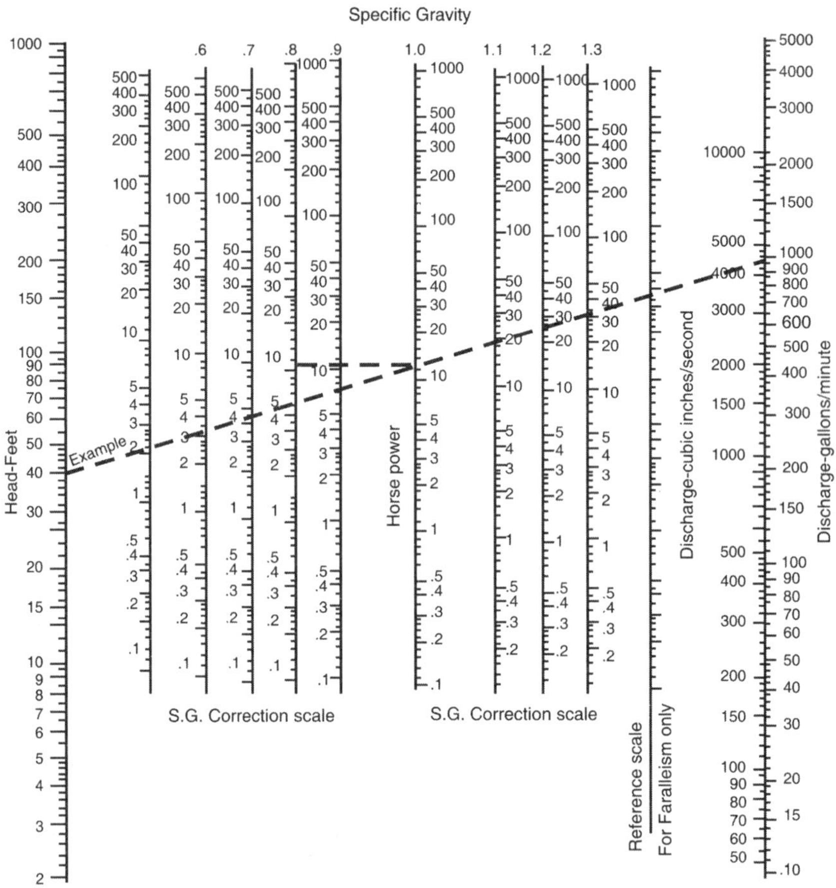

Fig. 1.34: Power output for given head, discharge and specific gravity chart

2 Pump Manufacture

2.1 POSITIVE DISPLACEMENT PUMPS

2.1.1 RECIPROCATING AND PLUNGER PUMPS

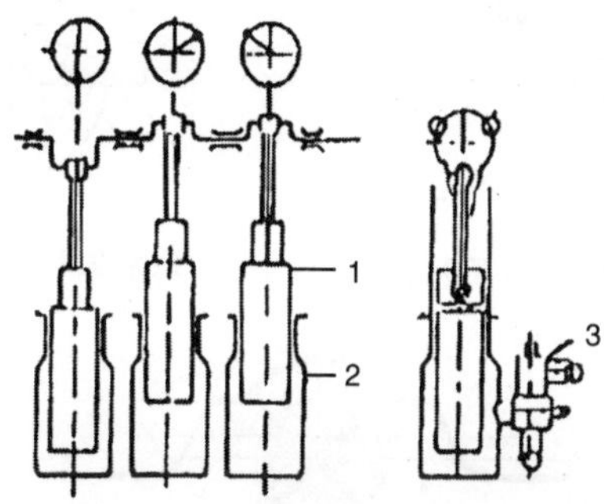

Fig. 2.1: Plunger pump

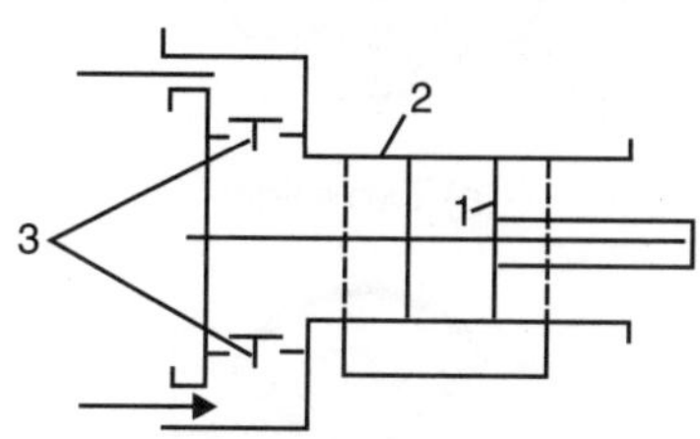

(1) Piston (2) Cylinder (3) Valves

Fig. 2.2: Piston pump

Figures 2.1 and 2.2 show the basic construction of one type of reciprocating and plunger pump. As the crank rotates, the piston moves between IDC and ODC. During its motion, it creates suction and delivery pressures alternatively. During the suction stroke, the cylinder is filled with the liquid and during pressure stroke the liquid is pushed through the delivery. The main features of a reciprocating pump are (*a*) positive displacement (*b*) slow speed (50 to 150 rpm) (*c*) self Priming and (*d*) high starting torque. These pumps are used mainly for feeding small boilers, gasoline pumping, fuel oil pumping, sprinkling, irrigation jobs etc., where high delivery pressure and low capacity conditions are required. Reciprocating pumps are either single acting or double acting. These pumps run in parallel with many pumps, operated with one crank shaft, to increase the quantity at the same pressure.

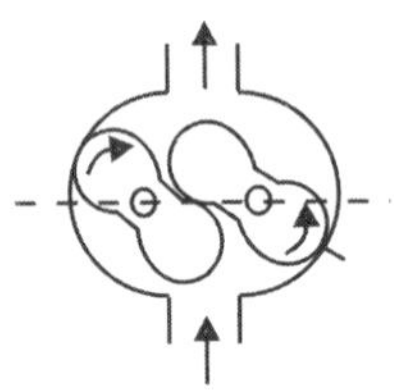

(1) Suction (2) Delivery (3) Lobe

Fig. 2.3: Lobe pump

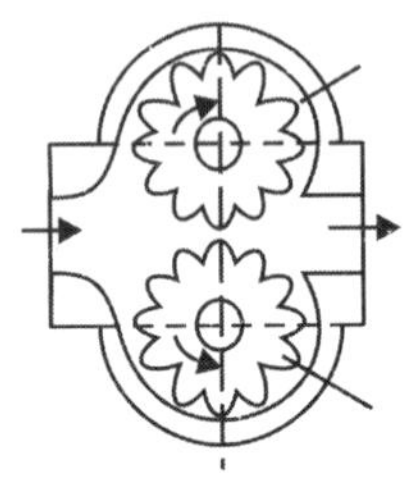

Fig. 2.4: Gear pump

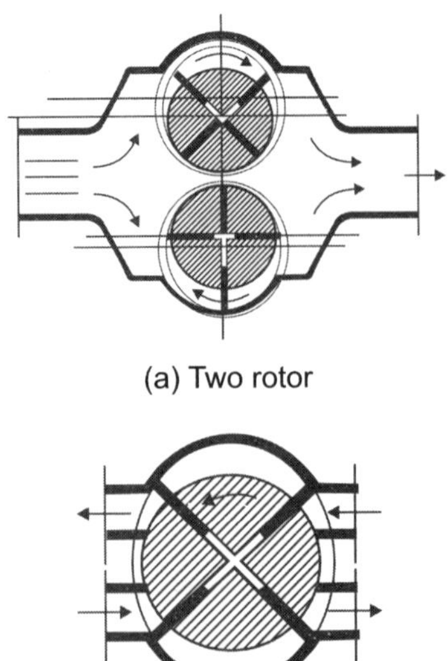

(a) Two rotor

(b) Single rotor two way

Fig. 2.5: Vane pump

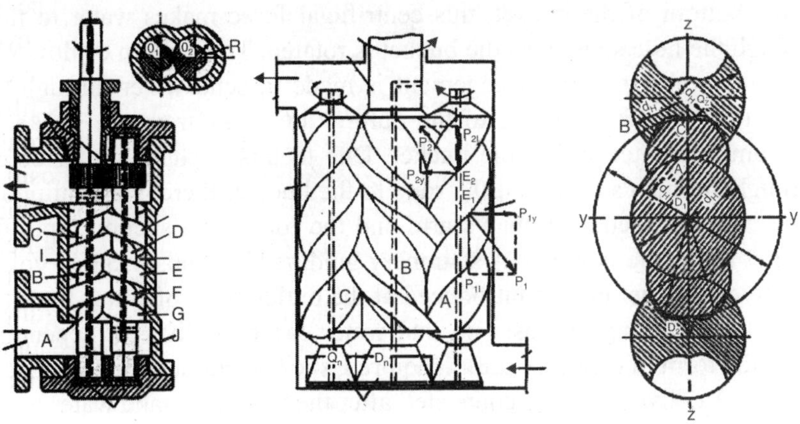

(a) Two screws (b) Three screws

Fig. 2.6: Screw pump

2.1.2 Rotary Pumps

A variety of rotary pumps are available, such as, Gear pumps (internal and external), Vane pumps (single vane, multi vane, sliding type, eccentric rotor type), Lobe pumps (two lobes, three lobes), Metering pumps or dosage pumps, radial piston and axial piston pumps etc. These pumps are used mostly for controlled smaller amount of flow conditions. The main features of rotary positive displacement pumps are (*a*) positive displacement (*b*) slow speed (*c*) self priming (*d*) fairly constant discharge (*e*) less vibration and (*f*) weight per unit flow is lower, when compared to reciprocating type and (*g*) lower hydraulic and mechanical friction because of less number of parts in contact with each other, (*h*) higher efficiency and (*i*) high starting torque.

2.2 ROTO DYNAMIC PUMPS

Basically all roto dynamic pumps are operated by centrifugal force. Hence they are called centrifugal pumps. A centrifugal pump operates by the force exerted by an object, moving in a circular path. Figure 2.7 illustrates the elementary theory of centrifugal pumps. A bucket is rotated by a person about his own axis, standing at the same place. The bucket whirls along with the man and gradually whirls abreast in a horizontal direction. Even in that condition, water does not spill out of the bucket, however fast, the man rotates. This force is the “Centrifugal Force”. If a hole is made at the bottom of the bucket, this centrifugal force makes water to flow through the hole so long as the bucket is rotated. This stream of flow will be continuous, if some arrangement is made to send water through the rope, to the bucket. Owing to the outflow of water from the bucket, a vacuum is created inside the bucket. Due to this vacuum, water enters through the man's arms and the rope to the bucket, thereby a continuous flow is established. The man's arm and the rope form suction pipe; the bucket forms the impeller. A stationery guiding mechanism called volute casing or diffuser can be introduced at the outlet point of water from the bucket for guiding purposes so that water can be delivered, depending upon the form of energy at outlet required, such as pressure or potential or kinetic. A delivery pipe is connected after the volute, to take water to the required point.

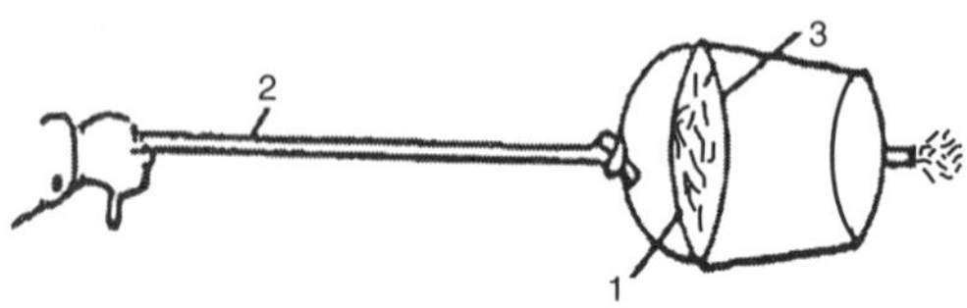

(1) Bucket (2) Water (3) Rope

Fig. 2.7: Principles of centrifugal pump

2.3 CONSTRUCTION OF A CENTRIFUGAL PUMP

A centrifugal pump essentially consists of an impeller with vanes, surrounded invariably by a volute casing. The impeller is mounted firmly on shaft, which is supported by two supports. One ball bearing and another bush bearing. Both these supports are kept in a housing called bearing bracket or bearing housing. A coupling is fitted at the other end of the shaft to get the drive from the prime mover such as electric motor or IC engine or turbine. The impeller rotates with close tolerance inside the casing.

The casing has connections for suction and delivery pipes. Liquid enters from the suction pipe to the impeller at the centre called **eye of the impeller**. Due to the rotation of the impeller by the prime mover, a mechanical force is imparted on the liquid by the impeller vane, there by the liquid flows towards the outer periphery. While flowing, the mechanical energy from the shaft is converted to hydraulic energy at the impeller and hence the liquid leaves impeller with higher hydraulic energy. This energy consists of three forms – potential energy, kinetic energy and pressure energy. The purpose of the casing is to convert these three forms of energy into any other form of hydraulic energy which is required at the end use. A delivery pipe connected at the outlet of the casing takes away the liquid to the place where it is used. Following Figs. 2.8 to 2.12 illustrate different roto dynamic pumps.

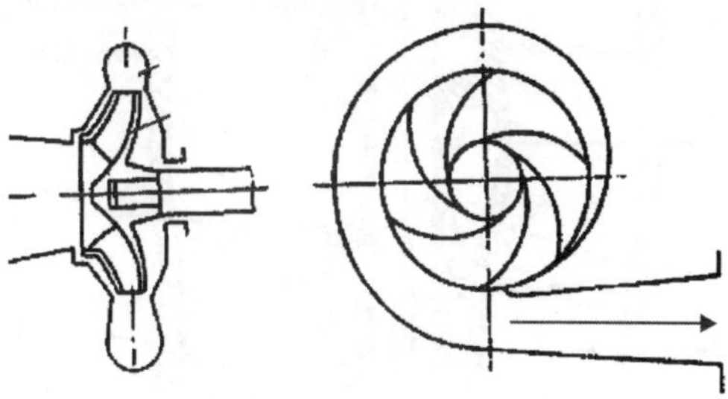

Fig. 2.8: Radial flow pump (Volute type)

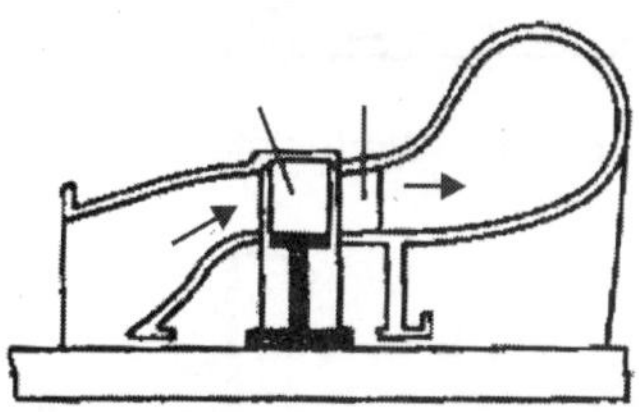

Fig. 2.9: Axial flow pump (Volute type)

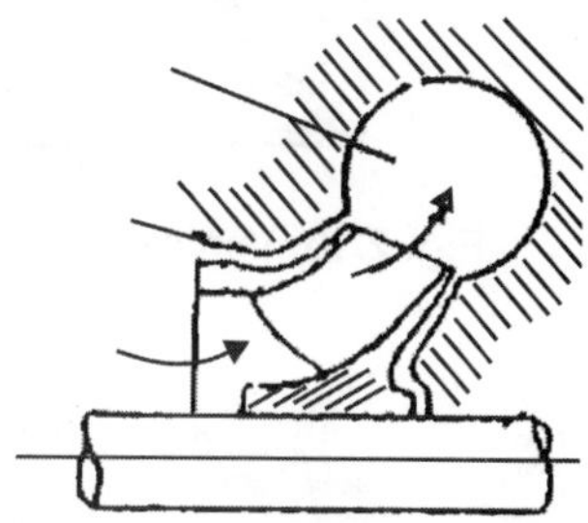

Fig. 2.10: Mixed flow pump (Volute type)

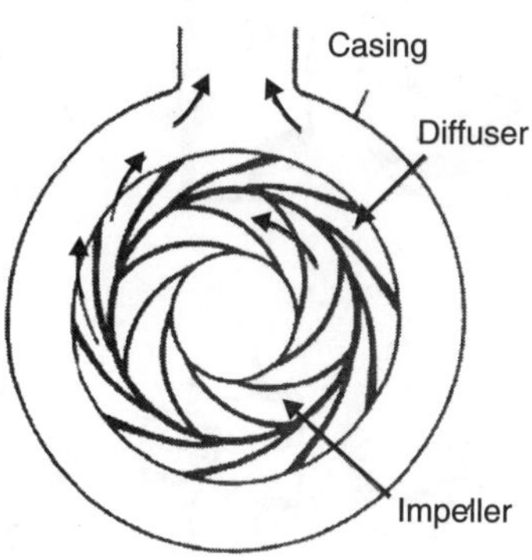

Fig. 2.11: Radial flow pump (Diffuser type)

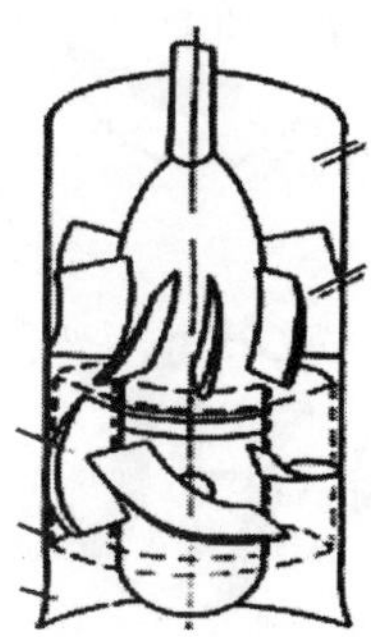

Fig. 2.12: Axial flow pump (Diffuser type)

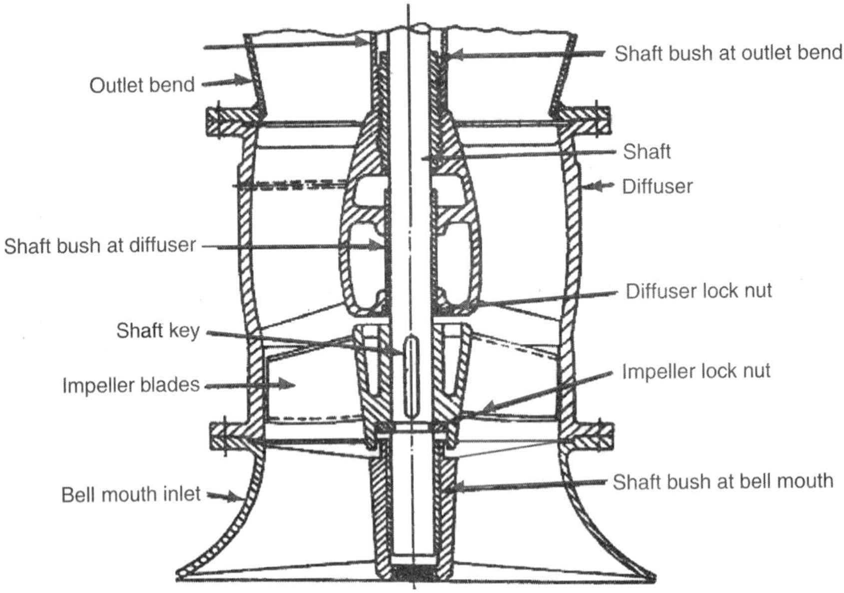

Fig. 2.13: Vertical wet-pet diffuser pump (Worthington pump)

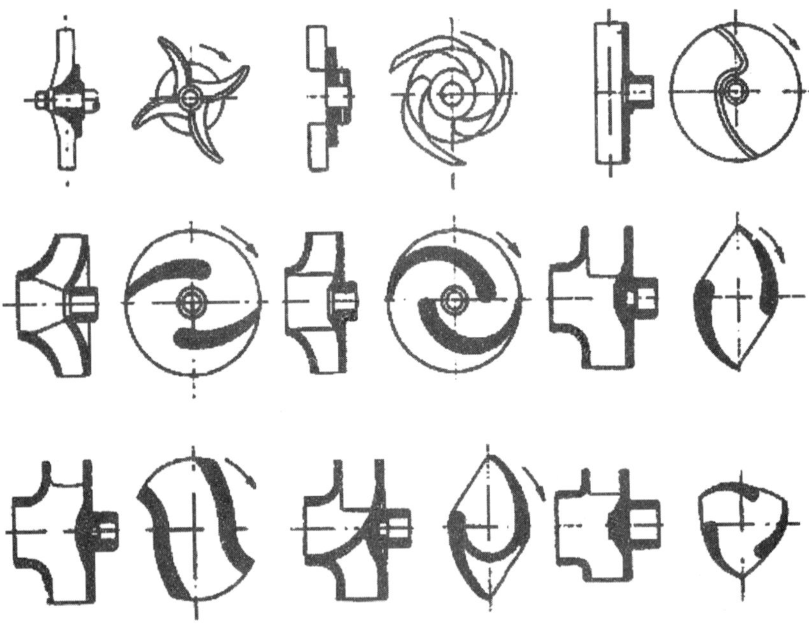

Fig. 2.14: Impeller shapes for high viscous and consistency liquids

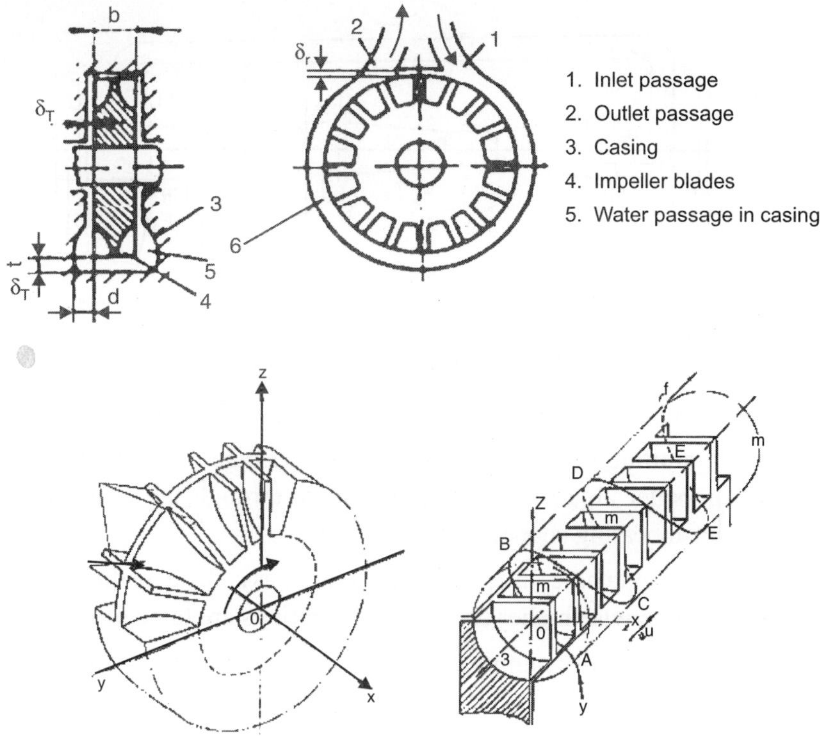

Fig. 2.15: Regenerative pump

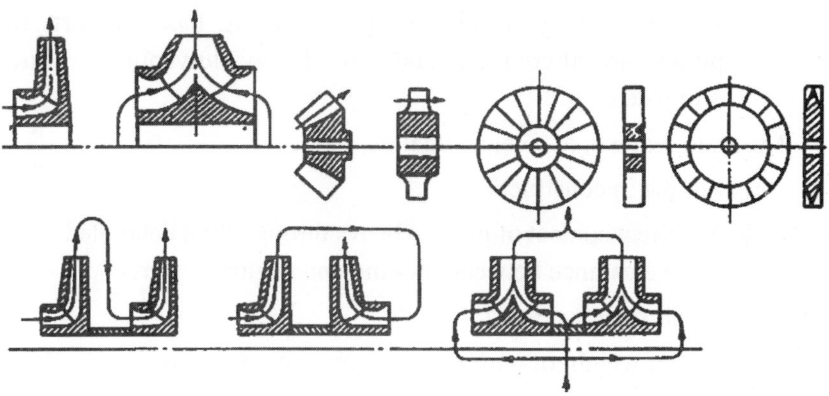

Fig. 2.16: Impellers for water and for low density viscous liquids

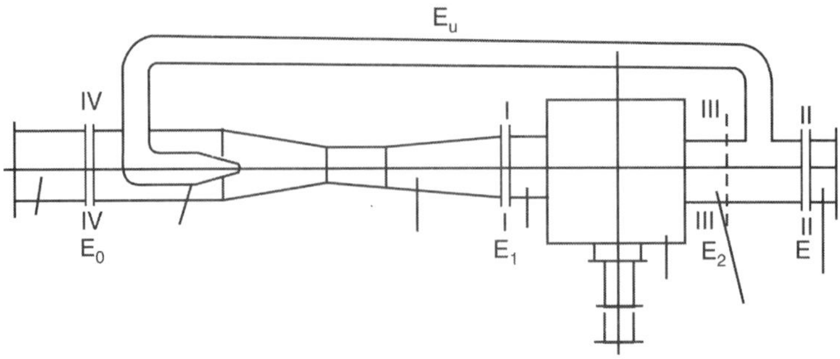

Fig. 2.17: Jet pump with centrifugal pump

Some more arrangements, provided in centrifugal pumps are:

(*a*) A stuffing box is provided to avoid leakage of the liquid from impeller area to outside atmosphere through the clearance in bush bearing support.

(*b*) Balancing holes are provided to reduce axial load on the bearing, as otherwise higher capacity ball bearings are to be provided.

(*c*) In heavy duty pumps or in high speed pumps two ball bearing supports, along with bush bearing support are provided to give more rigidity to the system. The entire pump and the prime mover are assembled over a common base with proper alignment, which enables to reduce the erection time at the site. According to the type of operations, base supports for pumps are altered especially for heavy duty pumps, slurry pumps etc.

The main features of a centrifugal pump are:

(*a*) Simple in construction,

(*b*) No direct contact of parts. Always through fluid entangled in the close clearance between moving and stationary parts,

(*c*) Small floor space,

(*d*) Lesser weight of pump per unit quantity pumped,

(*e*) Low initial and maintenance cost,

(*f*) Easy maintenance,

(*g*) Continuous flow,

(*h*) Quiet in operation,

(*i*) Less wear and tear of parts since there is no direct contact of parts and no valves or no reciprocating motion, and

(*j*) High speed of rotation. Centrifugal pumps are used in different forms. The following sketches show different ways of impeller assembly in pumps.

2.4 CENTRIFUGAL PUMP ASSEMBLY

Basically any centrifugal pump mainly consists of three major assemblies namely (*a*) casing assembly, (*b*) bearing bracket assembly, and (*c*) shaft assembly.

Casing assembly consists of casing, wear Plates, wearing Rings at suction end and at bracket end, and suction and delivery flanges. If there are wearing rings, these are fitted to the wear plates. In some of the constructions, such as back pull out types, suction and casing wear plates are made as one part. In some other designs such as non-clog pumps, the casing and bracket side wear plates are made as separate unit. The supporting leg for the pump is either given at the casing or at the bearing bracket according to the design criteria.

Bracket assembly consists of main bearing bracket, bearing covers, and intermediate brackets. In some designs, the intermediate bracket and main bracket are made as single unit. Stuffing box is also provided in some designs.

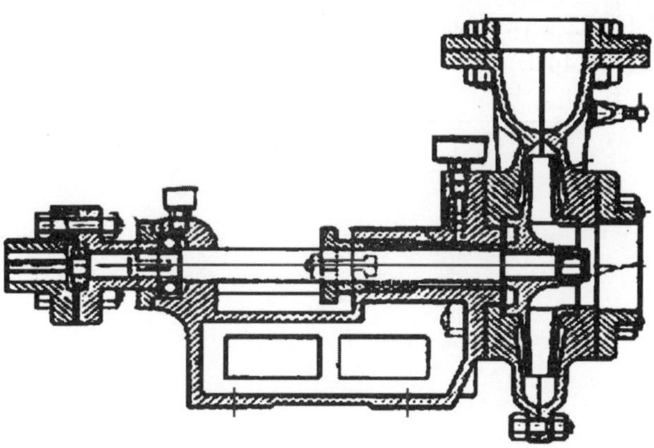

Fig. 2.18: Single ball bearing supported centrifugal pump

Shaft assembly consists of shaft, couplings (intermediate, pump and motor couplings), coupling nut, bearings, impeller and impeller nut. This assembly must be dynamically balanced for smooth and vibration free running. However, in practice, impeller alone is balanced, since all other parts are machined components (symmetrical) and hence balanced.

The following materials are applicable for normal executions. For liquids, having different temperatures and characteristics, other materials are recommended.

Table 2.1: Normal selection of materials for pumps

Pump part	Material	Pump part	Material
Spiral Casing	GG 26 CS – C 25 G – × 15 CrNiMo 18.9 G – × 10 CrNiMoNb 18.10 (1.4410)	Shaft	St 60.11 × 20 Cr 13 (4021 × 10) CrNiMoTi 18.10 (1.4571)
	G – × 10 CrNiMoNb 18.10 (1.4580)	Shaft Sleeve	G – × 70 CrMo 29.1 (1.4136) G – × 10 CrNiMoNb 18.10 (1.4580) Stellitiert
Intermediate Bearing	GG 26 G – × 15 CrNiMo 18.9 (1.4410)	Bearing Bracket Bantern	GG 22 G – × 15 CrNiMo 18.9 (1.4410)
Impeller	GG 26 G – × 15 CrNiMo 18.9 (1.4410)	Bearing Bracket	GG 22

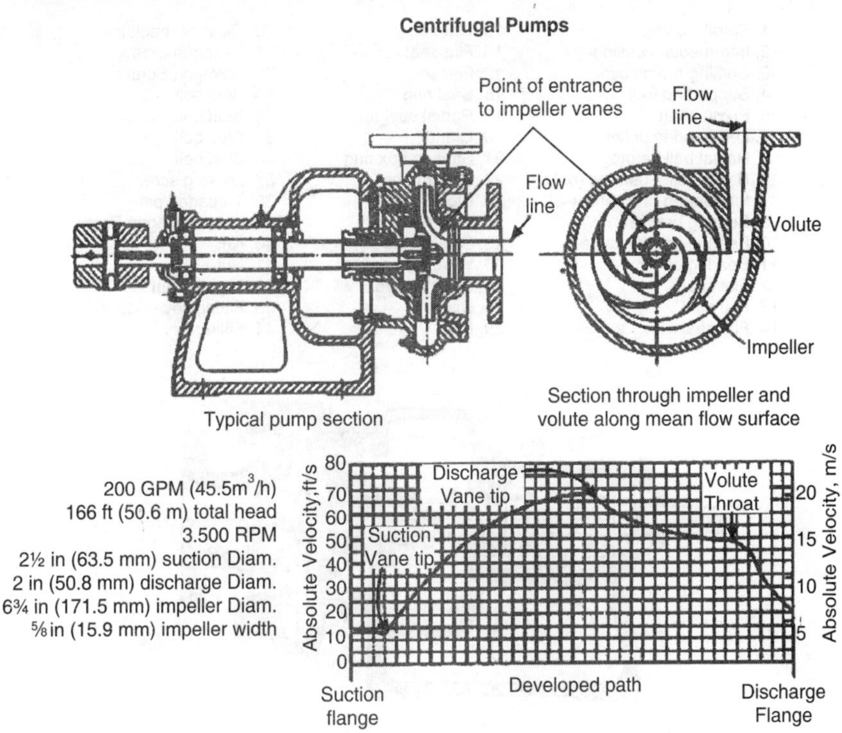

Fig. 2.19: Double ball bearing supported pump

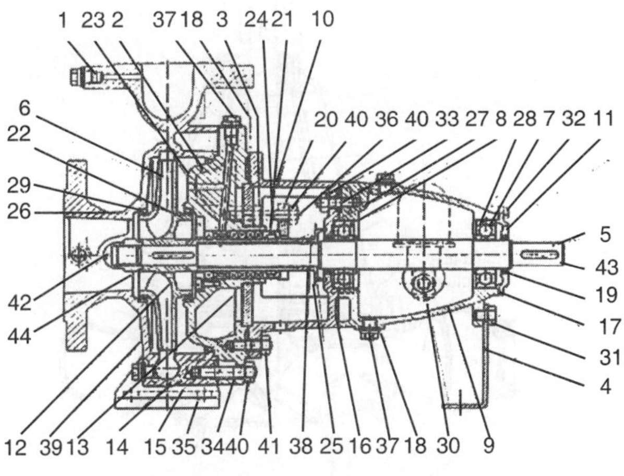

Fig. 2.20: Single stage end suction volute pump, double bearing supported (Worthington)

1. Spiral Casing
2. Intermediate casing
3. Cooling room cover
4. Supporting foot
5. Pump shaft
6. Left-hand-impeller
7. Radial ball bearing
8. Radial roller bearing (only for bearing bracket 2 + 3)
9. Bearing bracket
10. Bearing bracket lantern
11. Bearing cover
12. Flat seal
13. Flat seal
14. Flat seal
15. Flat seal
16. Flat seal
17. Flat seal
18. Seal ring
19. Radial seal ring
20. Gland
21. Stuffing box ring
22. Bottom ring
23. Block ring
24. Stuffing box
25. Splash ring
26. Wearing ring
27. Shaft sleeve
28. Disk
29. Pin
30. Oil level regulator
31. Hexagon Screw
32. Hexagon Screw
33. Stud bolt
34. Stud bolt
35. Stud bolt
36. Stud bolt
37. Locking screw
38. Threaded pin
39. Inner Hexagon Screw
40. Nut
41. Nut
42. Impeller nut
43. Fitting key
44. Fitting key

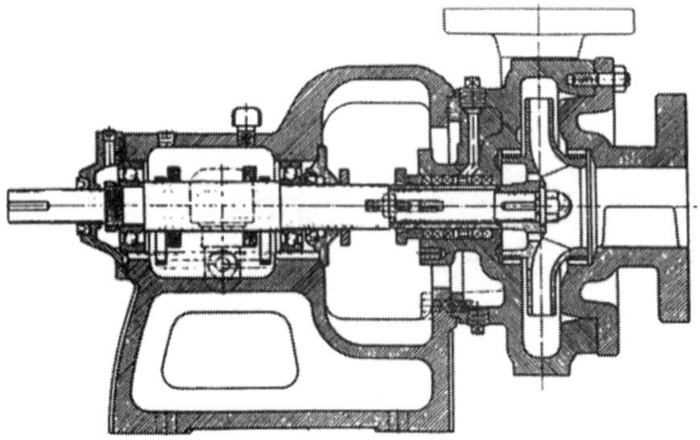

Fig. 2.21: Ball bearing pump with oil rings

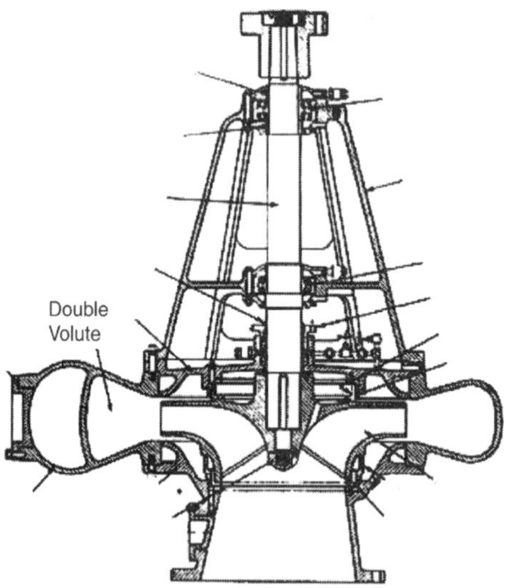

Fig. 2.22: Sectional view of a vertical-shaft end-suction pump with a double-volute casing.

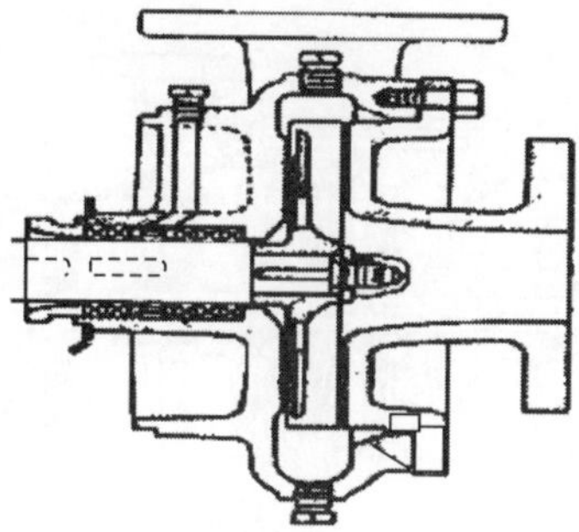

Fig. 2.23: End-suction pump with open impeller and removable suction cover (Worthington pump)

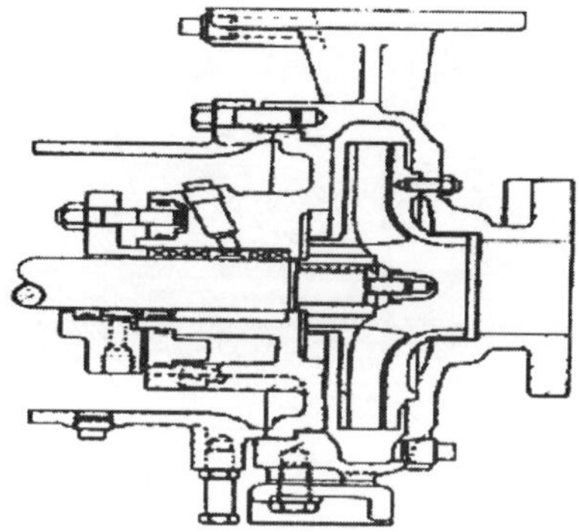

Fig. 2.24: End-suction pump with semi open impeller and renewable side plate (Worthington pump)

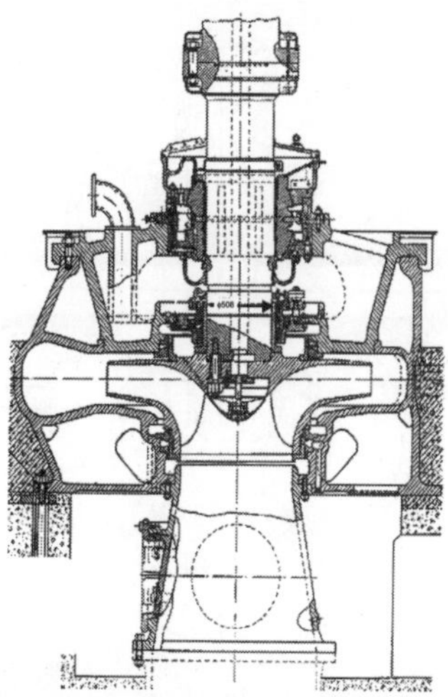

Fig. 2.25: Single volute vertical pump

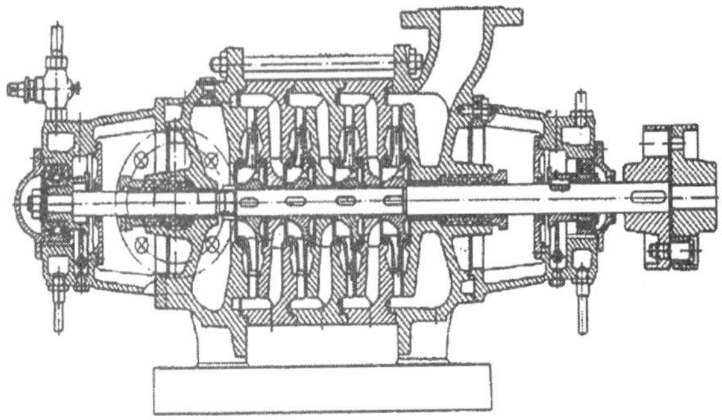

Fig. 2.26: Multistage horizontal pump with in line impellers

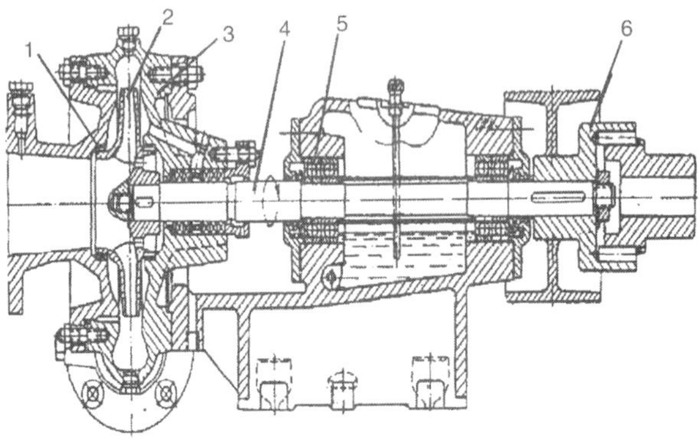

Fig. 2.27: Pump for heavy duty viscous liquid

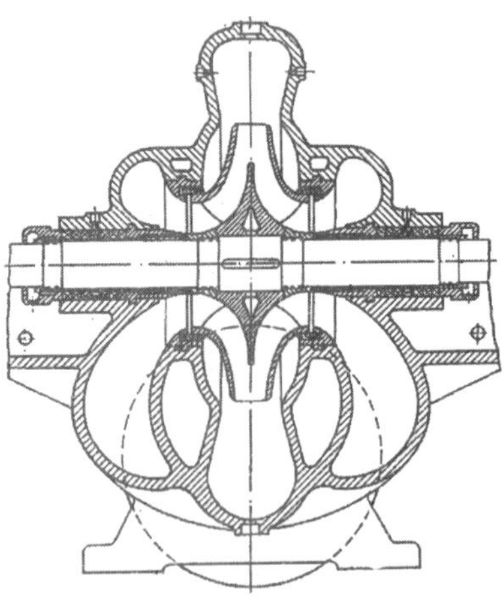

Fig. 2.28: Horizontal single stage double suction pump

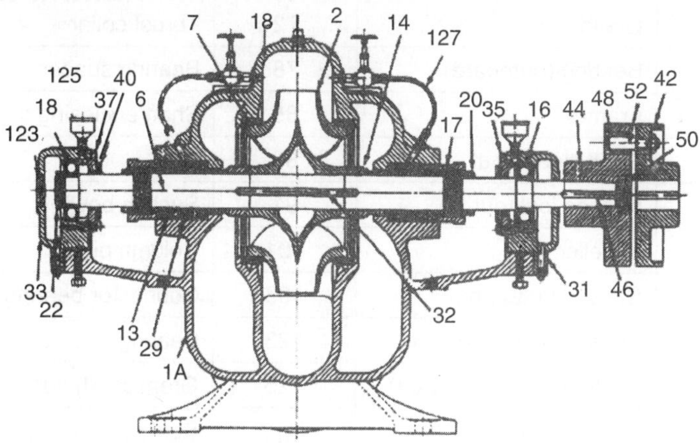

Fig. 2.29: Horizontal single-stage double-suction volute pump. (Numbers refer to parts listed in Table 2.2.) (Worthington pump)

Table. 2.2: Names of centrifugal pumps parts

Item no.	Name of part	Item no.	Name of part
1	Casing	33	Bearing housing (outboard)
1A	Casing (lower half)	35	Bearing Cover (inboard)
1B	Casing (upper half)	36	Propeller key
2	Impeller	37	Bearing cover (outboard)
4	Propeller	39	Bearing bushing
6	Pump shaft	40	Deflector
7	Casing ring	42	Coupling (driver half)
8	Impeller ring	44	Coupling (pump half)
9	Suction cover	46	Coupling key
11	Stuffing box cover	48	Coupling bushing
13	Packing	50	Coupling lock nut
14	Shaft sleeve	52	Coupling pin
15	Discharge bowl	59	Handhold cover
16	Bearing (inboard)	68	Shaft colliar

17	Gland	72	Thrust collar
18	Bearing (outboard)	78	Bearing spacer
19	Frame	85	Shaft enclosing tube
20	Shaft sleeve nut	89	Seal
22	bearing lock out	91	Suction bowl
24	Impeller nut	101	Column pipe
25	Suction head ring	103	Connector bearing
27	Stuffing box cover ring	123	Bearing end cover
29	Lantern ring (seal cage)	125	Grease (oil) cup
31	Bearing housing (inboard)	127	Seal pipe (tubing)
32	Impeller key		

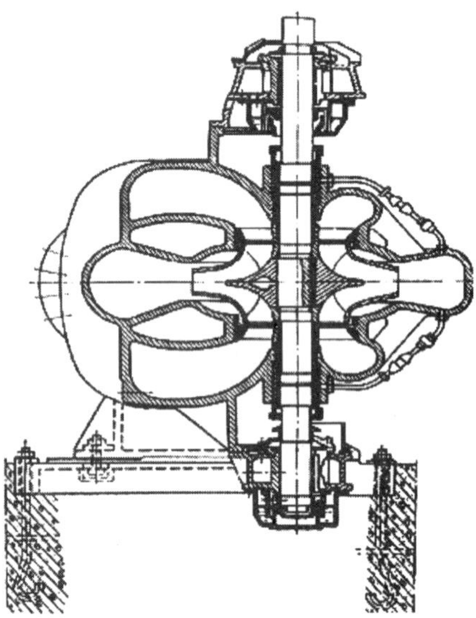

Fig. 2.30: Vertical single stage double suction pump

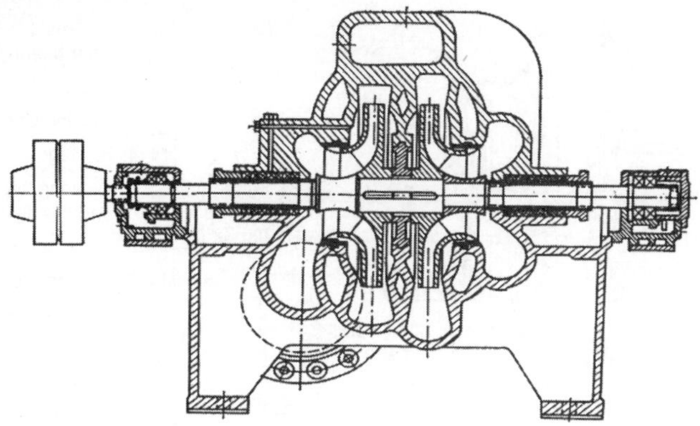

Fig. 2.31: Two stage horizontal volute type with opposed impellers

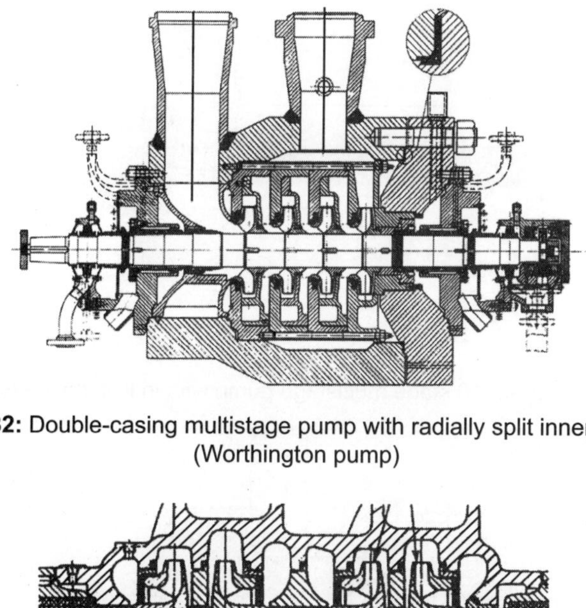

Fig. 2.32: Double-casing multistage pump with radially split inner casing. (Worthington pump)

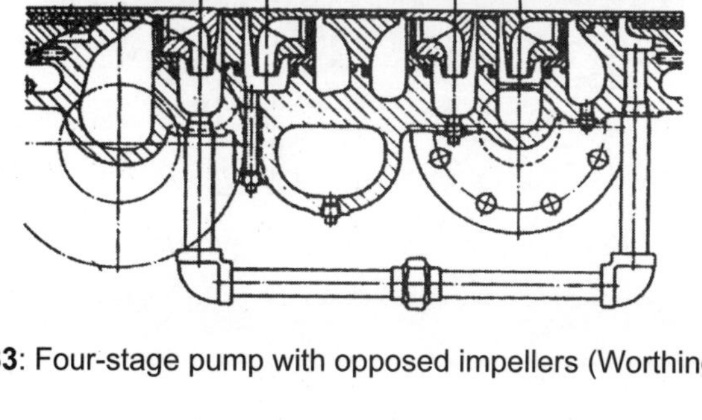

Fig. 2.33: Four-stage pump with opposed impellers (Worthington pump)

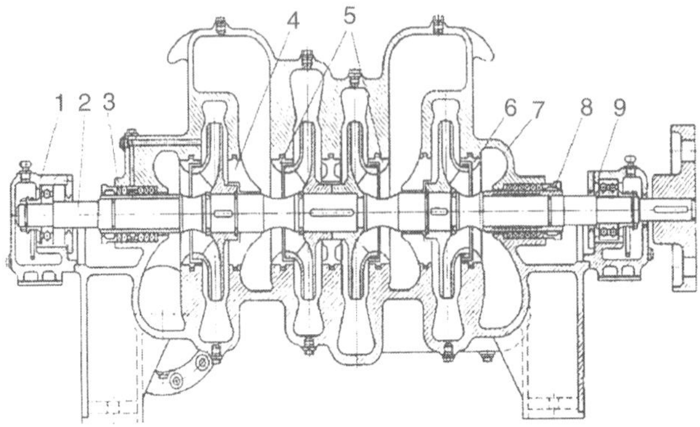

1. Rear bearing
2. Shaft
3. Stuffing box
4. Ist stage rear bracket
5. Wearing rings (impeller and casing)
6. Impeller - Ist stage
7. Suction volute
8. Stuffing box-gland
9. Front bearing

Fig. 2.34: Four-stage opposed impellers pump

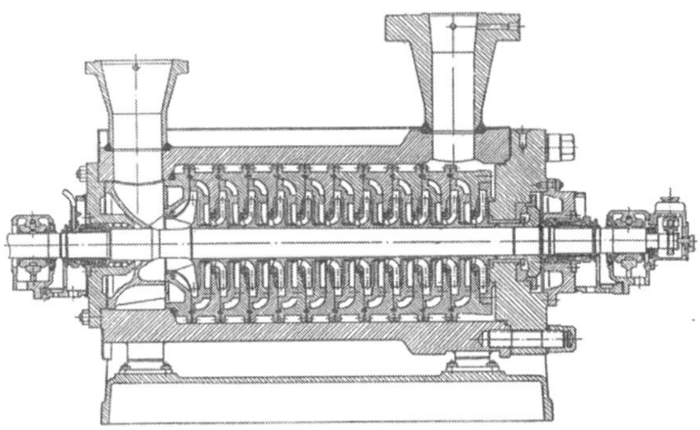

Fig. 2.35: 10 stage multistage pump with in line impellers

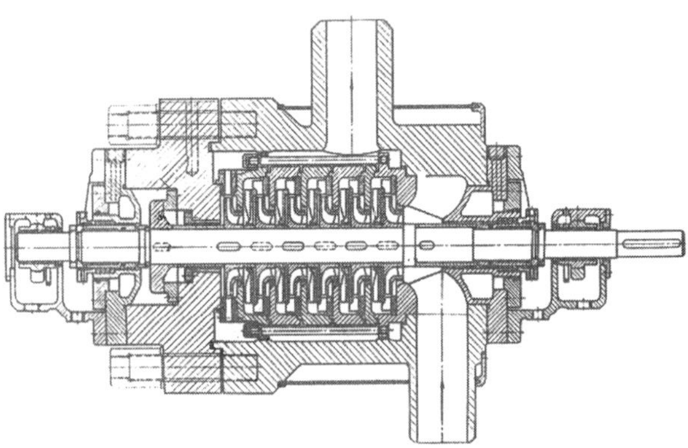

Fig. 2.36: Six stage in line impellers enclosed type

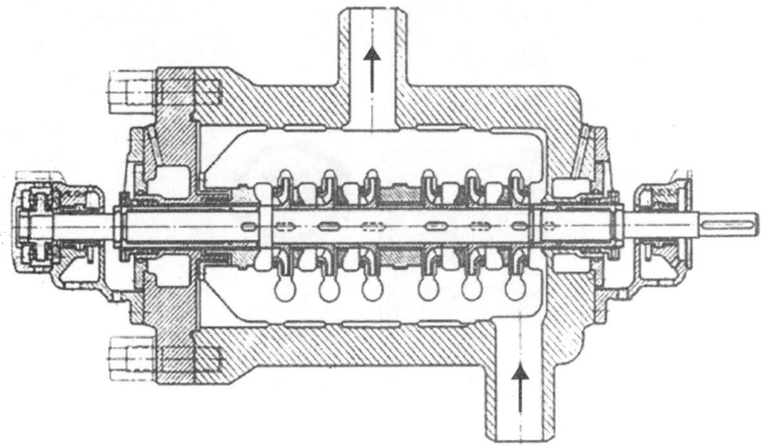

Fig. 2.37: Six stage opposed impellers console type

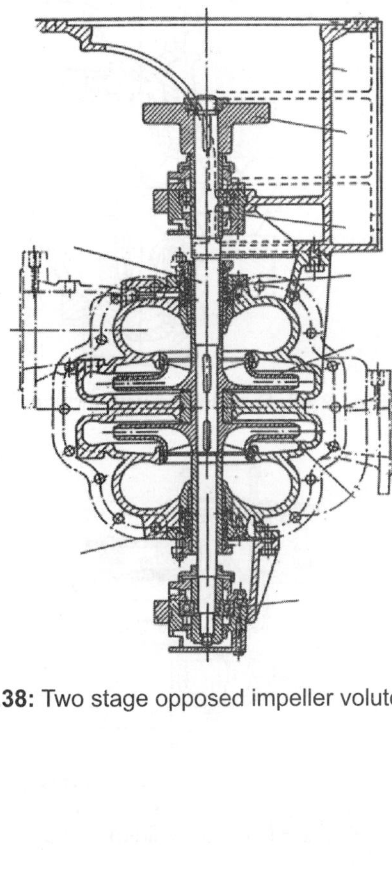

Fig. 2.38: Two stage opposed impeller volute casing

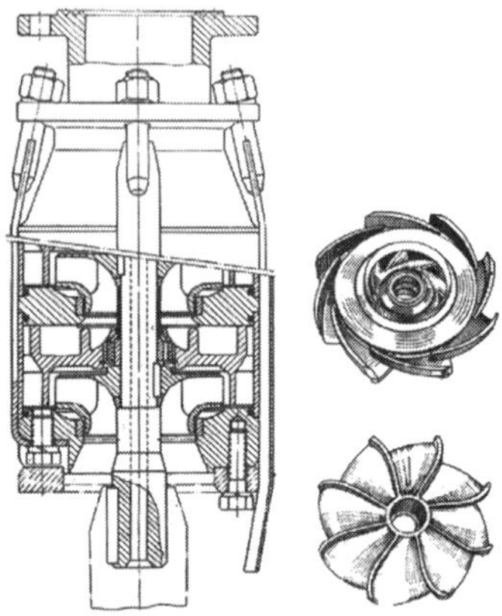

Fig. 2.39: Vertical diffuser type bore well pump

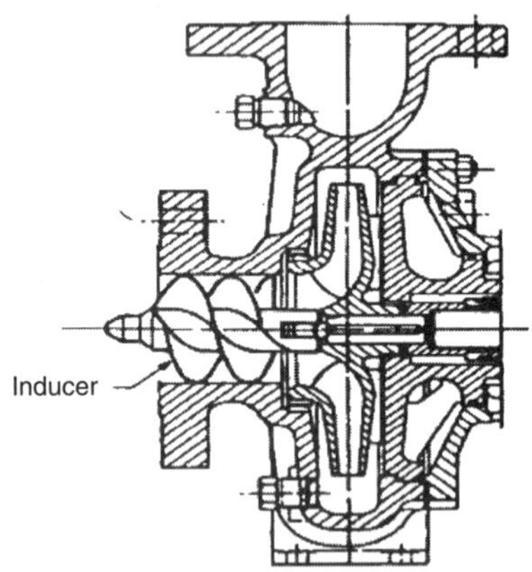

Fig. 2.40: Pump fitted with inducer.

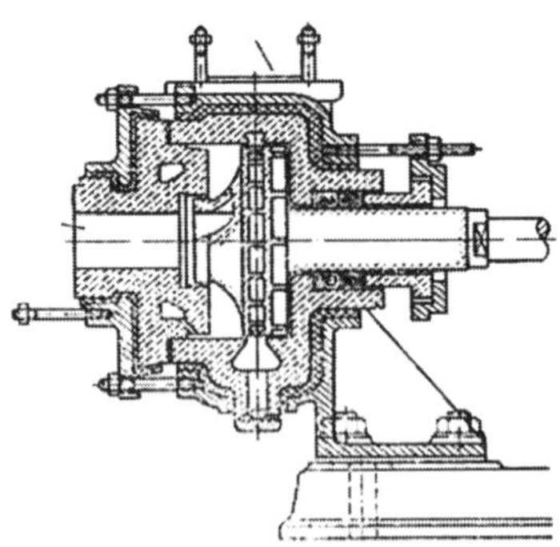

Fig. 2.41: Ceramic lined pump

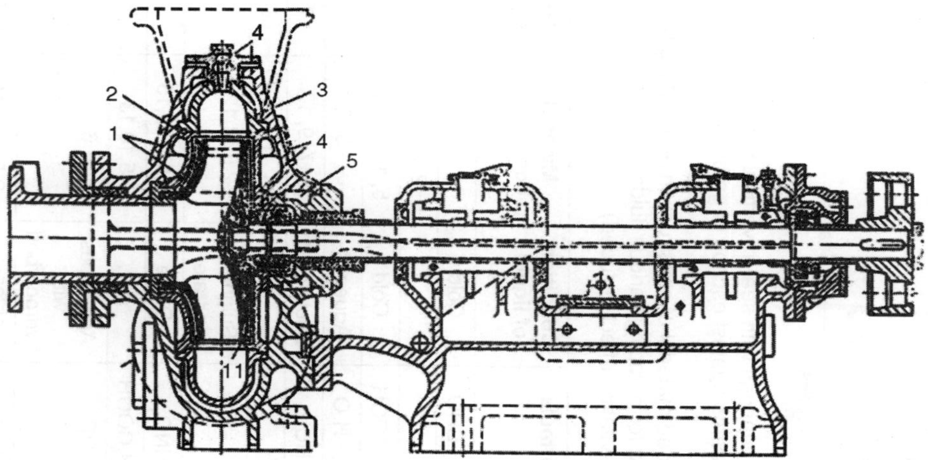

1. Casing linings, to suit the viscous liquid (impeller and casing)
2. Inner casing
3. Outer casing
4. Cooling passage for casing
5. Cooling passage for stuffing box

Fig. 2.42: Heavy duty high viscous lined pump

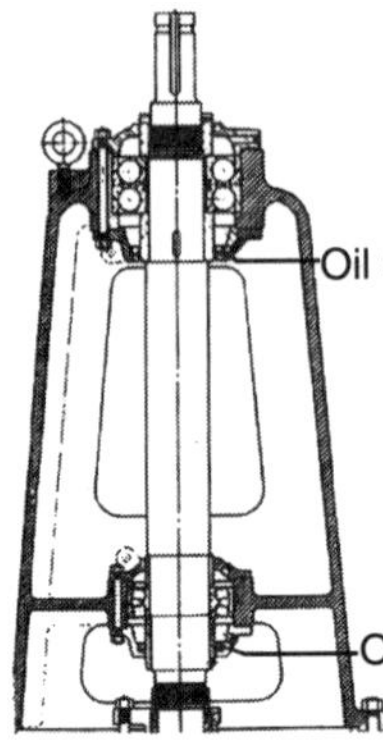

Fig. 2.43: Ball bearings with seal in vertical pump.
The seal guards against escape of grease

Table. 2.3: Recommended material of construction and stuffing box packing for pumping various fluids
(Cavitation effect included)

Sl. no.	Fluid	Formula	Condition of liquid	Material recommended	Packing selection number
1	Acetaldehyde	CH_3CHO	Cold	1	7
2	Acetate solvents	-		1, 2, 3,5,10,13	7
3	Acid, Acetic	CH_3COOH	Cold	5, 10, 13, 15	7, 10
4	Acid, carboric	$C_2 + H_2O$	Aqueous	3	7, 10
5	Acid, hydrochloric	HCL	Dilute cold	6, 7, 9, 12	7, 10
6	Acid, mime water	-	-	3, 5, 10, 11, 12	7, 10
7	Acid, mixed	-	Sulphuric nitrate	2,5,11,12	10
8	Acid, nitric	HNO_3	Dilute	4,5,7,12,15	10
9	Acid, benzoic	C_6H_5COOH	-	5,10,13	7, 10
10	Acid, hydrofluoric	HF	Anhydrous with hydrocarbon	6	10
11	Acid, sulphuric	H_2SO_4	65/93% 175 C	2,5,6,10,11	7, 10
12	-do- (oleum)	$H_2SO_4 + SO_3$	Fuming	-	10
13	Acid, Sulphuric	H_2SO_4	65/93% 175 C	2, 5, 6, 10, 11, 12, 13, 14, 15	7, 10
14	-do-	-do-	10 percent	5, 6, 10, 11, 12, 13, 14, 15	7

15	Acid, sulphurous	H_2SO_3	-	3, 5, 10, 13, 14	10
16	Acid, tan	-	-	3, 5, 6, 10, 13	7, 10
17	Alcohols	-	-	1, 3, 5	10
18	Alum., sulphate	$Al_2 (SO_4)_3$	Aqueous sol (pure)	5, 6, 11, 14, 15	5
19	Ammonia aqua	NH_4OH	-	2, 12	5
20	Ammonium chloride	NH_4Cl	Aqueous sol	5, 6, 11	5
21	Ammonium nitrate	NH_4NO_3	-do-	2, 5, 4, 6, 11	5
22	Ammonium phosphate, dibasic	$(NH_4)_2 HPO_4$	-do-	1, 2, 5, 6	3, 5
23	Ammonium sulphate	$(NH_4)_2 SO_4$	-do-	2, 4, 5, 11	5
24	Asphalts	-	-	1, 5	7
25	Barium nitrate	$Ba (NO_3)_2$	Aqueous sol	2, 5, 6, kk	-
26	Beer	-	-	3, 5	4
27	Beet juice	-	-	3, 5	4
28	Beet pulp	-	-	1, 2, 3, 5, 10, 11, 13	4
29	Benzene	C_6H_6	-	1,2,5	10
30	Benzol	-	-	1,2,5	10
31	Brine, alkai			2,5	-
32	Brine, Acidic			5,7	-
33	Brine, chilling			2	-
34	Butane	C_4H_{10}		1,2	10
35	Brine, sea water	-		1,2,3,4,5	4
36	Calcium chloride acqueous sol.	$CaCl_2$	PH_3	1	4
37	-do-	-do-	-do-	3,4,5,6	4
38	Calcium magnesium chloride	-	Aqueous	-do-	4
39	Calcium sodium chloride	-	-do-	-do-	4
40	Calcium hypochlorite	$Ca(OCl)_2$	-do-	1,5,6,15	4
41	Cane Juice	-	-	1,3,4	4
42	Carbon bisulphide	CS_2	-	2	3
43	Carbon tetrachloride	CCl_4		1,2,3,4,5	10
44	Chloro benzene	C_6H_5Cl		3,5,6,10,13	10

45	Chloroform	$CHCl_3$		3,5,6,10,13	10
46	Cellulose acetate	-	Aqueous	5,6	4
47	Copper ammonium acetate	-	Aqueous	1,5,10,13	4
48	Copper chloride	$CuCl_2$	Cupric aqueous	6,7,9,15	4
49	Enamel	-	-	1	4
50	Ethane	C_2H_6	-	1,2	19
51	Ethylene chloride (di-chloride)	$C_2H_4Cl_2$	Cold	3,5,6	10
52	Ferrous sulphate	$FeSO_4$	Aqueous	5,6,14,15	4
53	Fruit juices	-	-	3,5,6,10,3	2
54	Gasoline	-	-	1,2,5	3,10
55	Glycerol (Glycerine)	$C_2H_5(OH)_3$	-	1,2,3,5	3
56	Hexane	C_6H_{14}	-	1,2	10
57	Heptane	C_7H_{16}	-	1,2	10
58	Hydrogen perioxide	H_2O_2	Aqueous	5,10,13	10
59	Jar	-	Hot	1	-
60	Janning liquors	-	-	3,5,6,10,13,15	7
61	Lime water (milk of lime)	$Ca(OH)_2$	-	2,5,13	4
62	Liquor, pulp milk black	-	--	1,2,4,5,6,11	4,10
63	Liquor, pulp milk green	-	-	1,2,4,5,6,11	4,10
64	Liquor, pulp milk white	-		1,2,4,5,6,11	4
65	Liquor, pulp milk pink	-		1,2,4,5,6,kk	4,10
66	Lithium chloride	LiCl	Aqueous	1	-
67	Manganese chloride	$MnCl_2$	Aqueous	3,5,10,13,15	4
68	Milk	-	-	5	2,10
69	Molasses	-	-	1,3,5	4
70	Oil, cold tar	-	-	1,2,5,10,13	4,7
71	Oil, coconut	-	-	1,2,3,5,6,11	4
72	Oil, creosote	-	-	1,2	4
73	Oil, crude	-	Cold or Hot	1,2	4
74	Oil, kerosene	--	-	1,2,5	4,10

75	Oil, linseed	-	-	1,2,3,5,6,10,13	9
76	Oil, lubricating	-	-	1,2	9
77	Oil, palm	-	-	1,2,3,5,6,11	9
78	Oil, quenching	-	-	1,2	9
79	Oil, Palm	-		1,2,3,5,6,11	9
80	Oil, Rapessed	-		3,5,6,10,13	9
81	Oil, Soyabean	-		1,2,3,5,6,10,13	9
82	Oil, Turpentine	-		1,2	10
83	Oil,Parafin	-	Hot or Cold	1,2	4
84	Oil, petroleum ether	-	-	1,2,5	10
85	Pentane	C_5H_{12}	-	1,2	10
86	Potash	-	Plant liquor	3,4,5,6,11	6
87	Potash alum	$Al_2(SO_4)_3$	Aqueous	3,4,6,11	6
88	Potassium-carbonate	K_2CO_3	Aqueous	1,2	
89	Potassium- chlorate	$KClO_3$	Aqueous	5,10,13,15	6
90	Potassium Hydroxide	KOH	Aqueous	2,4,5,6,12	6
91	Potassium Nitrate	KNO_3	Aqueous	1,5,10,13	6
92	Propane	C_3H_8	Aqueous	1,2,	10
93	Pyridine	C_5H_5N		5	6
94	Sewage	-	-		9
95	Silver nitrate	$AgNO_3$	Aqueous		9
96	Slop, Brewery	-	-		4
97	Slop, distillers	-	-		4
98	Soap, liquor	-	-		4
99	Soda Ash	Na_2CO_3	Cold		4
100	Soda Ash	-do-	Hot		4
101	Sodium bicarbonate	$NaHCO_3$	Aqueous sol.		
102	Sodium chloride	NaCl	3.1% Cold		4
103	-do-	-do-	3.1% Cold		4
104	-do-	-do-	3.2% Cold		4, 10
105	Sodium hydroxide	NaOH	Aqueous sol.		5
106	Sodium silicate	Na_2SiO_3	-		4
107	Sodium Sulphate	Na_2SO_4	-		5
108	Starch	$(C_6H_{10}O_5)$	-		3

109	Sugar	-	Aqueous sol.		3
110	Vegetable juices		--		4
111	Vinegar		--		7
112	Water boiler feed		Not evaporated pH8.5		1
113	Water, high make up		pH8.5		9
114	Water, low make up		Evaporated		9
115	Water, chlorinated		--		1
116	Water, distilled		High purity		9
117	-do-		--		9
118	Water fresh		Condensate		9
119	-do-		Paper mills		4
120	Wood pulp		Stock		4
121	Zinc sulphate	$ZnSO_4$	Aqueous sol		7

Table. 2.4: Materials of construction

Sl. no.	Mateial of construction	Relevant specification
1	Bronze fitted*	Grade V of IS: 318-1962 +
2	All iron	Grade 20 of IS: 210-1962 +
3	All Bronze	Grade V of IS: 318-1962 +
4	All austentic iron	Type 2 od IS : 2749-1964
5	All stainless steel	Designation 07Crl9Ni9MoZ of IS: 1570-1961
6	All monel metal	Composition: Nickel - 63 percent Copper-30 percent Iron- 2 Percent Manganese- 0.9 percent Silicon- 4 percent Tensile strees-60kg/mm^2 Elongation on 5d- 20 to 10 % Brinell hardness-275 to 300
7	Rubber lined	
8	White iron	IS: 2107-1962

9	Gra/jote	
10	Hastelloy	Composition: Nickel - 85 percent Silicon - 10 percent Copper - 3 percent
11	Nickel cast iron	Composition: Nickel - 0.7 to 2%, Tensile stress- 18 kgs Iron- Balance, S.P. Gravity 7.3
12	Alumina porcelain	Electrical porcelain containing approximately 50% alumina (Al_2O_3)
13	Chlorimet 3	Composition: Nickel - 60 percent Chromium - 18 percent Molybdenum - 8 percent Iron, Silicon & Carbon - Balance
14	Antimonial lead (Heard lead)	Composition Lead - 94 percent Antiony - 6 percent
15	High silicon cast iron (duriorn)	Composition Silicon - 15 percent Iron, Carbon and Manganese-Balance
16	Ceramic (galss/ stoneware)	
17	PVG (unplastics zed)	
18	Polystyrene	
19	Glass lining	

- Bronze Fitted Pumps – The pumps in which the casing is of cast iron; the impeller, casing ring, the impeller ring and shaft sleeves are of bronze, and the shaft is of steel.
    - Specification for leaded tin bronze ingots and castings (revised).
    - Specification for grey iron castings (revised). Specification for austenetic iron castings. Schedules for wrought steels for general engineering purposes and specification for white heat malleable iron castings have not yet been formulated.

2.5 GASKETS, SEALS AND PACKINGS

The gaskets, seals and packings, used in special purpose pumps, must be suitably selected, so as to withstand the effect of liquid being pumped. This shall be selected generally in consultation with the manufacturer taking into account the application. Wherever possible, suitable mechanical seals are preferred to packings. The recommended materials for seals for special purpose pumps may be selected from one or more of the following. Seals, recommended for various types of fluids handled are given in Table 2.5. Wherever possible, mechanical seals should be preferred to those given in the table.

Table. 2.5: Seal selection

Sl. no.	Brief description of seals
1	Asbestos plaited yard seal greased, graphited and lubricated.
2	Asbestos plaited throughout yarn seal unlubricated.
3	Asbestos steam seal greased, graphited and lubricated.
4	Asbestos lubricated hard seal.
5	Asbestos rubberized laminated seal.
6	White metal foil seal crinkled lubericated asbestos yarn.
7	Blue asbestos anti-acid seal non-metallic.
8	Polytetrafluoroethylene (Teflon).
9	Cotton yarn (lubricated) seal.
10	Mechanical seal.

2.6 LEAKAGE LOSS

Leakage of flow takes place at different places inside the pump (1) the clearance between the impeller and casing at the suction side (2) the clearance between the impeller and casing at the bracket side of the impeller (3) flow through the balancing hole at the impeller and (4) flow through balancing disc to equalize the axial thrust. All these tend to reduce the actual quantity of flow of the pump. Volumetric efficiency and correspondingly the overall efficiency reduce. Different forms of clearance passage between the impeller and the casing are listed in the Figs. 2.44 to 2.56. Selection of these, vary according to the type of pump and pumping liquid. For high pressure units with gases, more complicated formation are adopted. However it should be noted that more complicated forms give rise to more difficult machining operations.

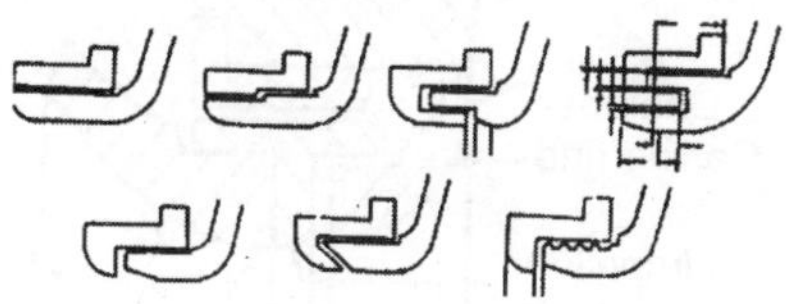

Fig. 2.44: Forms of clearance between wearing rings

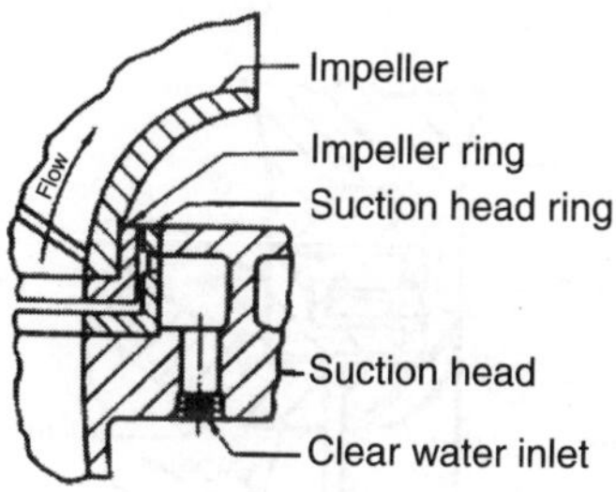

Fig. 2.45: Water-flushed wearing ring

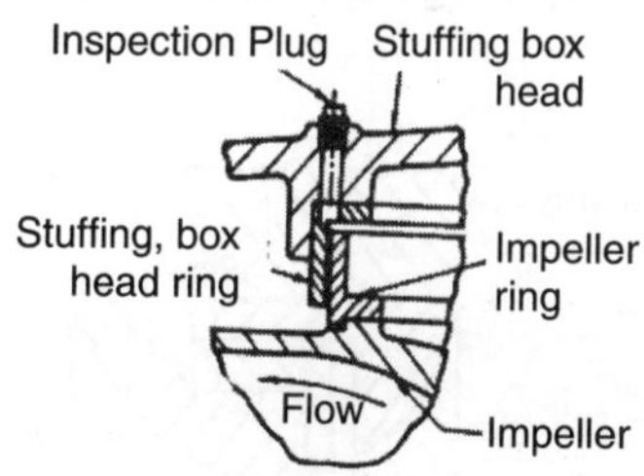

Fig. 2.46: Wearing ring with inspection port to check clearance

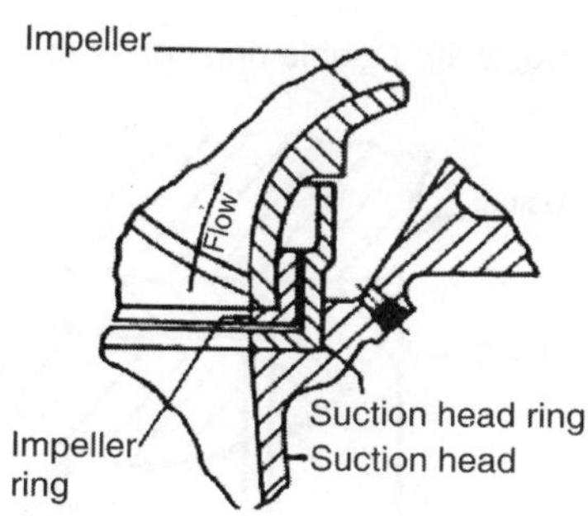

Fig. 2.47: Dam type ring construction

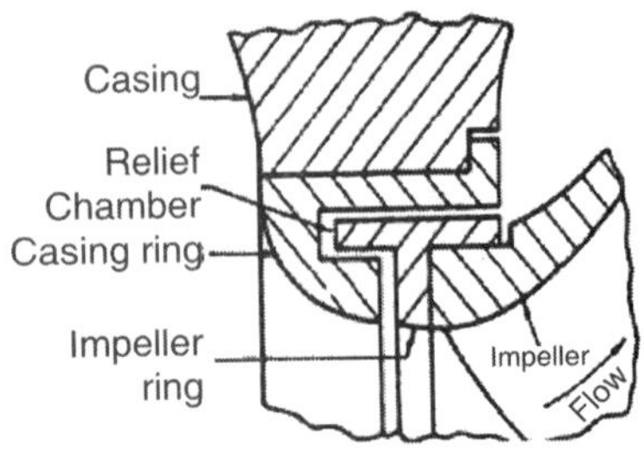

Fig. 2.48: Double ring type construction nozzle type casing ring

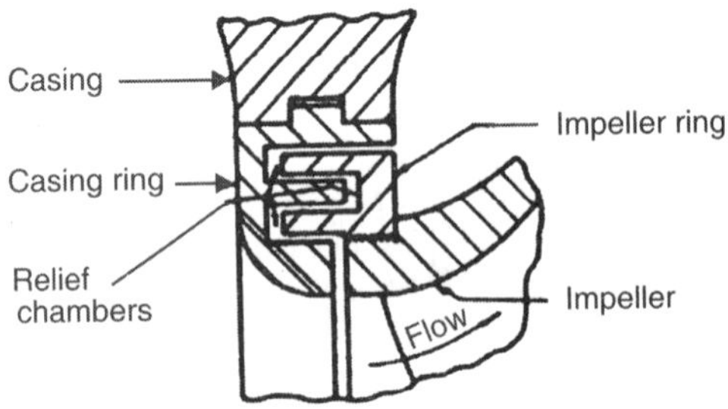

Fig. 2. 49: Labyrinth type rings in double ring construction, both type

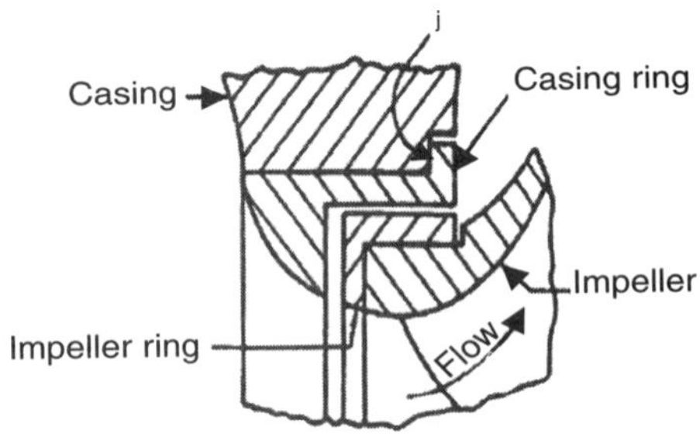

Fig. 2.50: Double ring construction

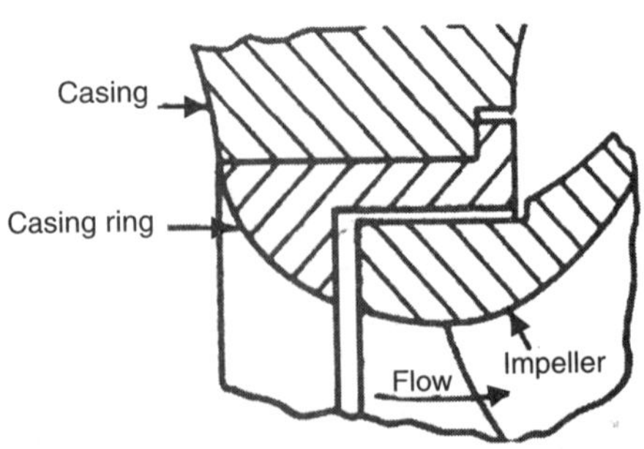

Fig. 2.51: Step type leakage joint with double rings

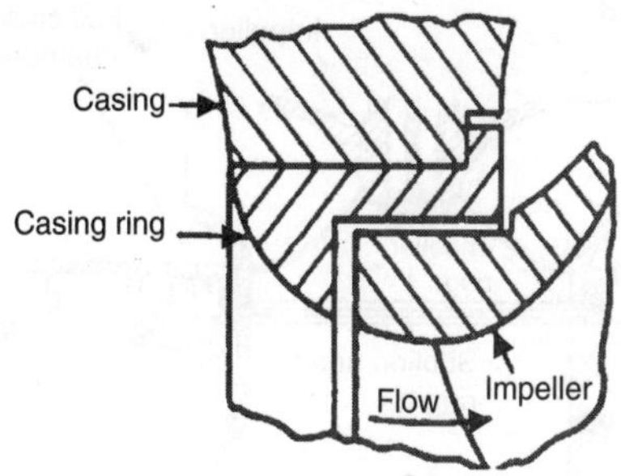

Fig. 2.52: An L type zozzle casing

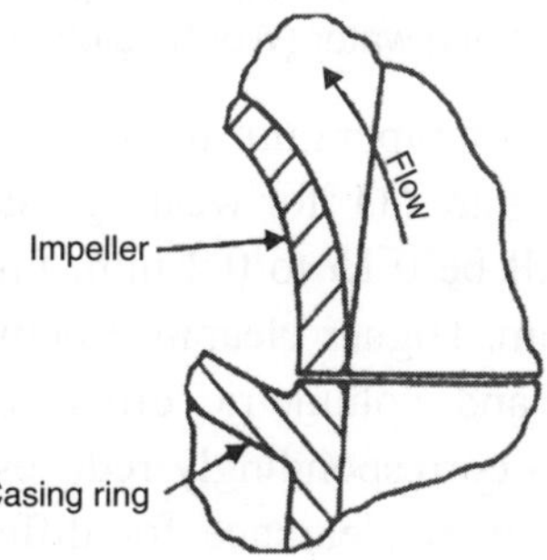

Fig. 2.53: Leakage joint with clearance

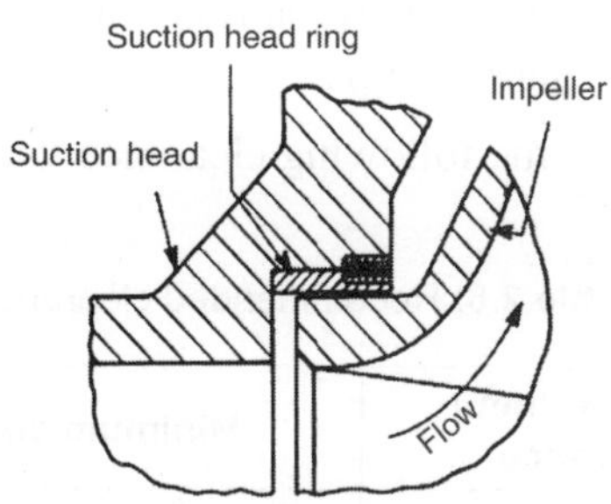

Fig. 2.54: Single flat casing ring construction

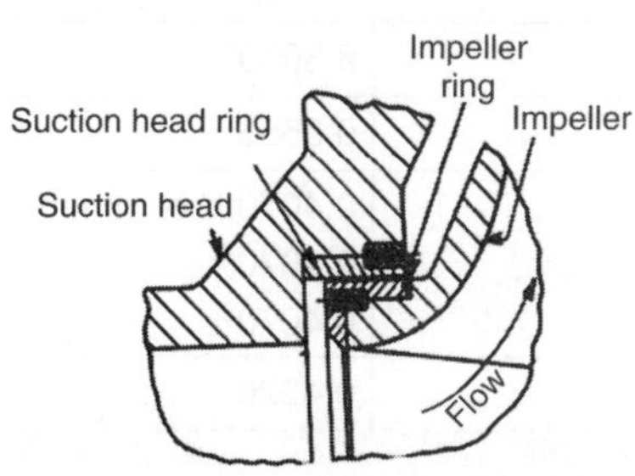

Fig. 2.55: Double flat ring construction

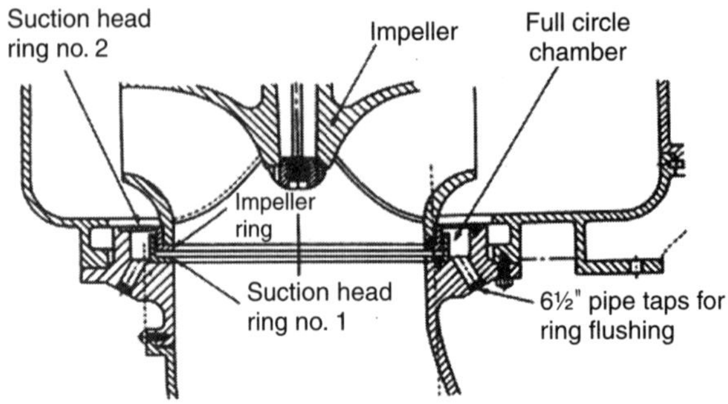

Fig. 2.56: Two sets of rings with space between for flushing water (Worthington pump)

For water at normal temperature the clearance between the rotating and stationary members are (1) for wearing ring diameter of rotor up to 100 mm, clearance will be 0.15 to 0.2 mm. For size above 100 mm, it is increased to 0.25 mm. Higher clearance between impeller and casing reduce the discharge and volumetric efficiency reduces considerably. Overall efficiency also correspondingly reduces. Figures 2.57, 2.58 and 2.59 show the variation of clearance for different wearing ring sizes, bushes and for positive displacement pumps.

Recommended clearance between rotary and stationary members for positive displacement rotary pumps.

In refinery practice, the following clearance values are recommended (Table 2.6).

Table 2.6: Recommended clearances

Diameter of rotating (mm) member at clearance	Minimum clearance (mm) dia
Less than 51 mm	0.2540
51 to 63.4 mm	0.2795
63.5 to 76.19 mm	0.3050
76.2 to 88.8 mm	0.3555
88.9 to 126.9 mm	0.4070 recommended for API 650 0.4070 0.4070
127 to 152.3	0.4320

152.4 to 177.7	0.4580
177.8 to 202.9	0.4820
203.0 to 228.4	0.5080
228.5 to 253.9	0.5340
254 to 279.4	0.5590
279.5 to 304.9	0.5850
305–	0.59
For materials having greater galling tendencies add 0.0127 mm to the aforementioned clearance.	

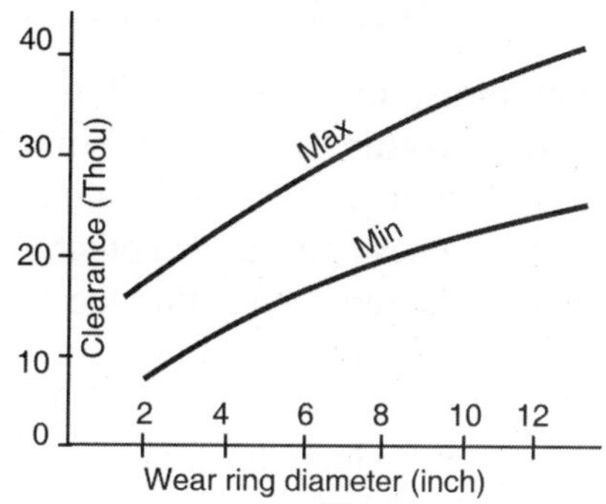

(1)

Fig. 2.57

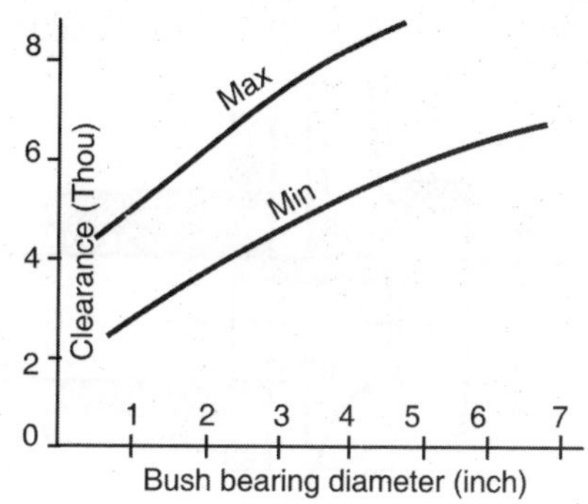

(2)

Fig. 2.58

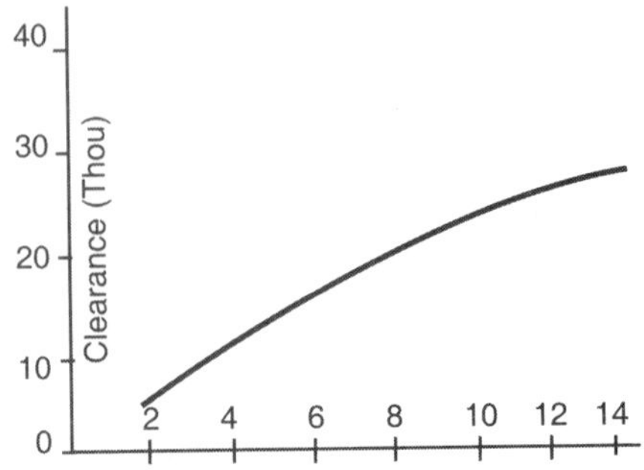

Fig. 2.59

2.7 AXIAL THRUST AND BALANCING DEVICES

The presence of axial thrust in pumps, is illustrated with a sketch considering Fig. 2.61, (*a*), pressure drop is uniform on both sides of the impeller when flow passes through the inner and outer shrouds up to the suction pipe size. Both the forces oppose each other. As a result of which, net axial force on impeller is zero. After this the pressure reduces still further at the rear end of the impeller and is positive, (above atmosphere), whereas pressure at the suction side is negative (under vacuum), as a result of which a net force acts on the impeller from the shaft. The magnitude of the force depends upon the delivery pressure, the suction pressure, the

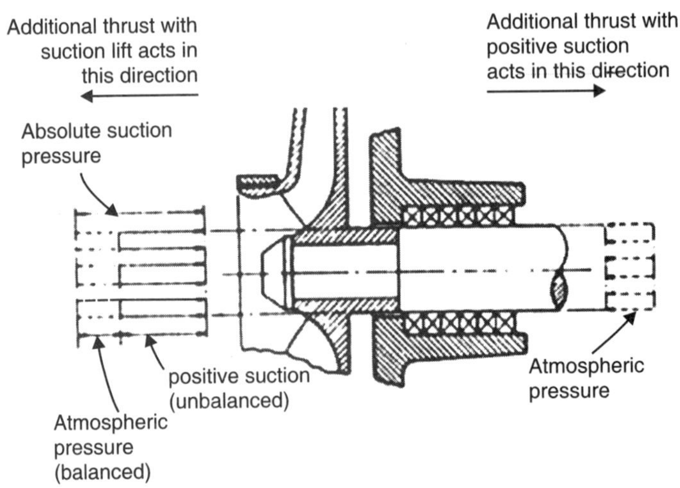

Fig. 2.60: Axial thrust problem with single suction overhung impeller and stuffing box

type of the clearance and the pressure drop at the clearance. Whatever may be the construction, axial force always exists.

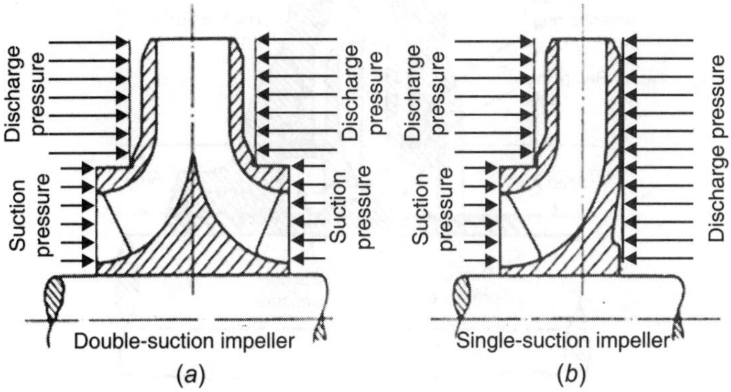

Fig. 2.61: Origin of pressures acting on impeller shrouds to produce axial thrust

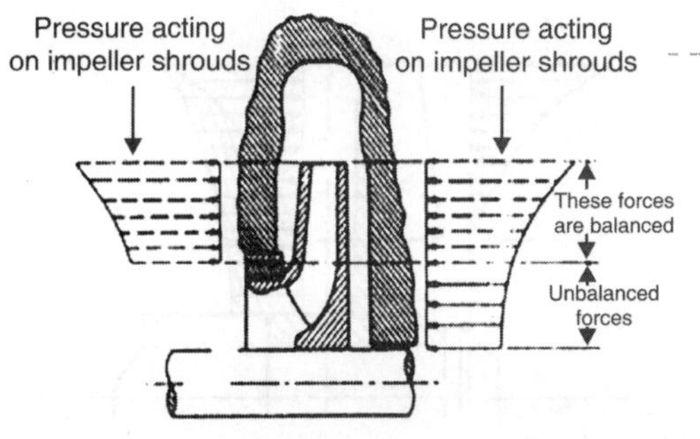

Fig. 2.62: Pressure distribution on front and back shrouds of a single suction impeller with shaft through impeller eye.

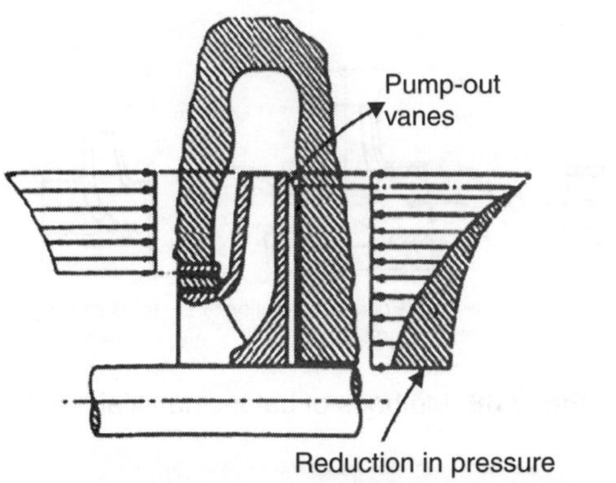

Fig. 2.63: Pump out vanes used in a single impeller to reduce axial thrust

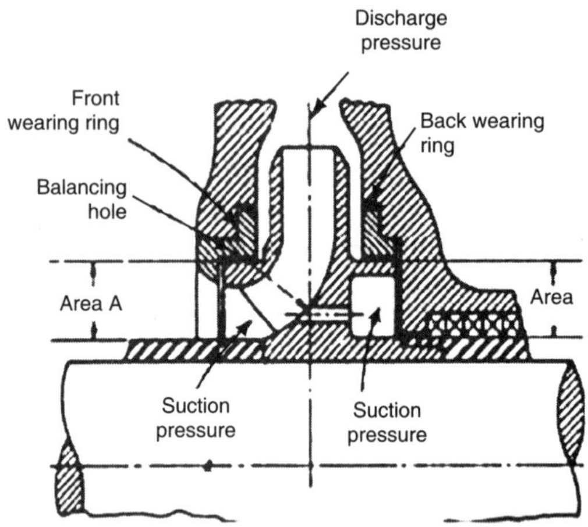

Fig. 2.64: Balancing axial thrust of single suction impeller by means of wearing ring on the back side and with balancing holes

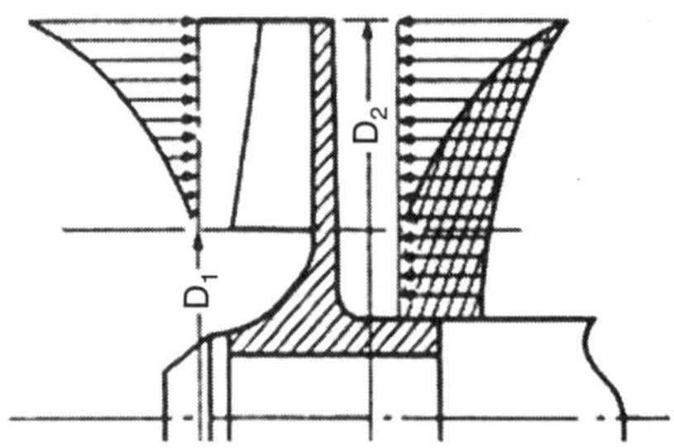

Fig. 2.65: Axial thrust in a semi open, single suction impeller

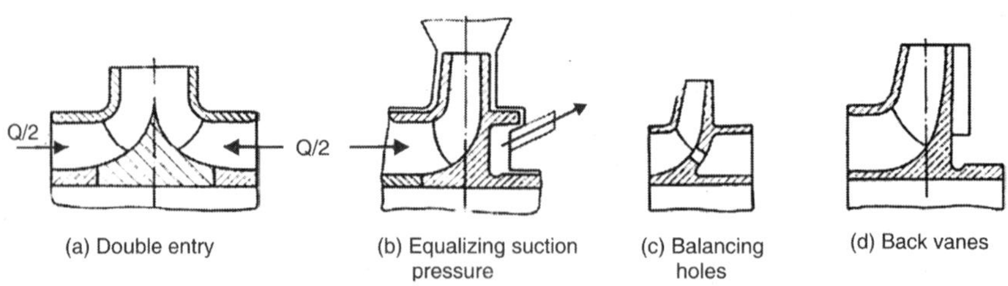

Fig. 2.66: Methods of balancing axial thrust

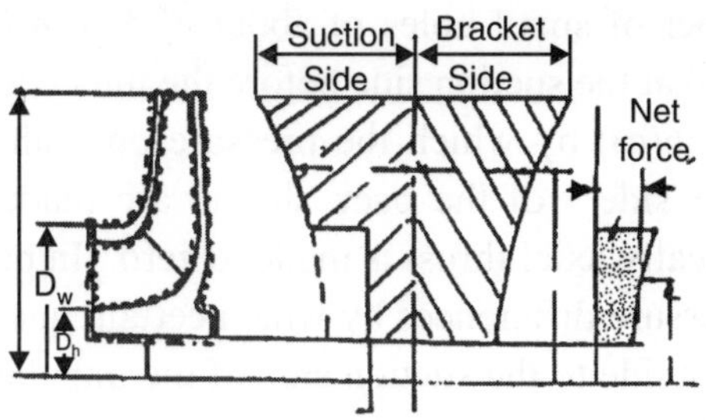

Fig. 2.67: Without back vanes

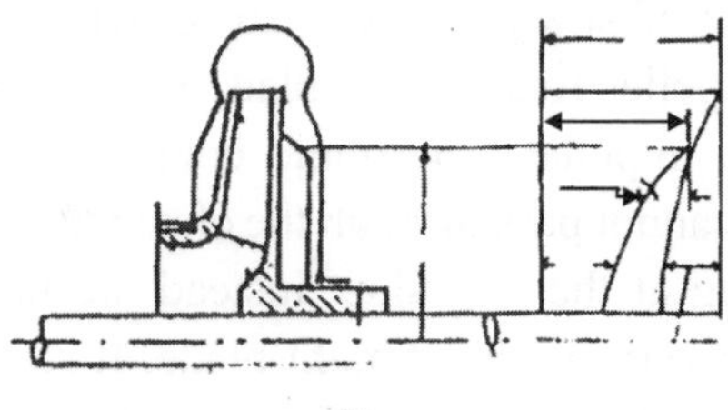

Fig. 2.68: With back vanes axial thrust

2.7.1 METHODS OF REDUCTION

Axial force is normally taken care of by the bearings provided at the bearing bracket. For smaller pumps, deep groove ball bearings and angular contact bearings are selected to take care of the axial thrust, but in larger single stage pumps, additional thrust bearings are provided.

For very high value of axial thrust, the bearing size becomes abnormally large, which is not only costlier but also the pump weight becomes unnecessarily higher due to the increase in the size of the bearing housing. In order to avoid axial thrust for single stage pumps, different balancing devices are adopted.

1. The impeller is designed as double suction type *i.e.*, liquid from the suction pipe enters both sides of the impeller, there by axial thrust is balanced (Figs. 2.28 to 2.30). In practice, perfect balancing cannot be attained due to unequal flow rate through each side of the impeller, but the magnitude of the axial thrust is very small which is taken care of by bearings. The direction of axial thrust changes due to the change of flow rate. Hence bearings at both ends must be designed for axial thrust.

2. A number of small holes of about of 5 to 8 mm diameter are provided at the suction side before the inlet edge of the impeller (Fig. 2.66(c)) by which the pressure on both sides (*i.e.*, front and rear side) of the back shroud are made equal, there by theoretically axial thrust is made to zero. In practice, there will be a pressure differences by which certain quantity passes from the outer side to the suction side of the impeller, which reduces actual flow coming out of the pump. Because of this there will be a little amount of axial thrust, which is taken care of bearings.
3. A number of straight blades are provided at the rear shroud. When impeller rotates, this blade system also rotates and acts as a mini impeller. The liquid coming out from the outlet of impeller cannot pass through the clearance between the impeller and casing at the rear side. Instead the liquid is pushed back to the delivery by the straight blade. It reduces the pressure at the bottom of the rear end, there by, the axial thrust is reduced. However a little amount of axial thrust exists, which is taken care of by the bearings. However power consumption increases due to extra power required to rotate the back vanes. In practice, a small quantity passes through the rear blades towards stuffing box. Due to this flow rate also reduces slightly.
4. A pipe is attached between the rear side of the impeller, and the suction sides of the impeller, there by the pressure at both sides are made equal theoretically. Practically there will be a slight pressure difference and hence slight axial thrust prevails, which is taken care of by bearings. Certain quantity passes through the pipe from rear side to the suction side, which reduces the flow rate of the pump. In all three cases (2), (3) and (4) flow rate is reduced, due to internal circulation of liquid. This reduces the volumetric efficiency and correspondingly overall efficiency. In case of back vanes, volumetric efficiency reduces, and the power consumption also increases. Hence, over all efficiency reduction will be higher than that of (2) and (4). In case of double suction impellers, the pump weight increases considerably. The weight per unit quantity will be larger in these pumps, when compared with other single stage pumps.

In multistage pumps, axial thrust balancing is done by a special device called "balancing discs and balancing drums". One such arrangement is shown in Fig. 2.69. The delivery pressure of the last stage and suction pressure are connected to a balancing chamber kept at the rear side of the last stage. The pressure difference causes a flow through the clearance between rotating and stationary balancing discs. Since axial force depends on this pressure difference and the same pressure difference exists at the balancing chamber also, these two forces equalize each other and hence axial thrust is zero.

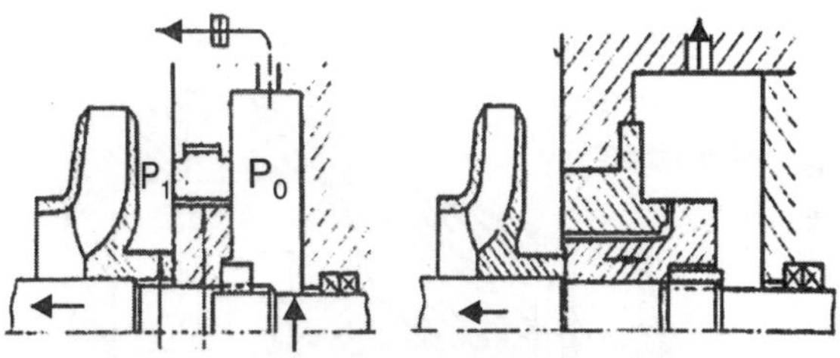

Fig. 2.69: Two Types of balancing methods

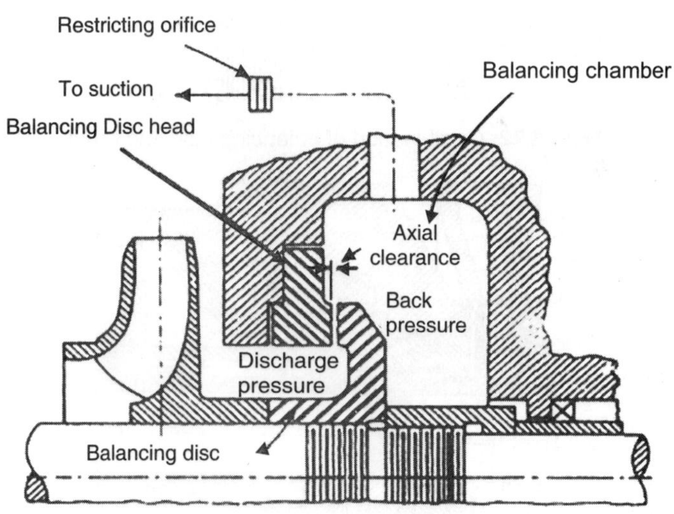

Fig. 2.70: Simple balancing disc

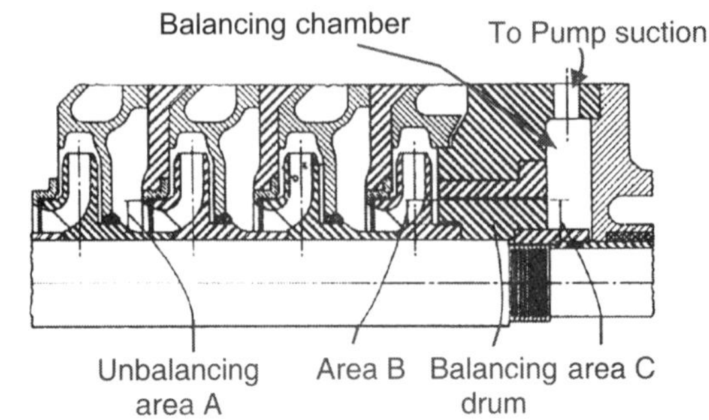

Fig. 2.71: Balancing drum

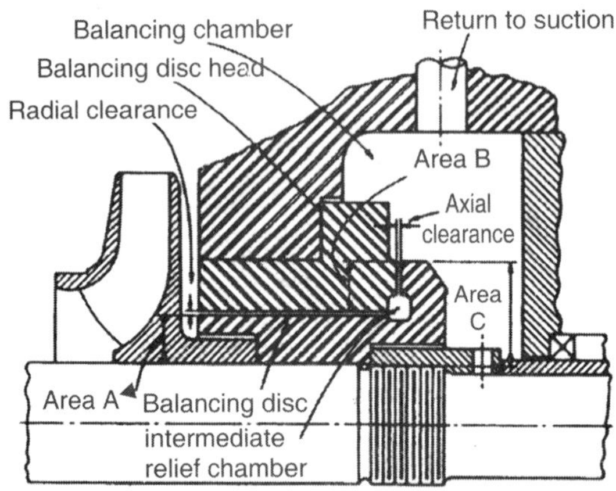

Fig. 2.72: Combination of balancing disc and drum

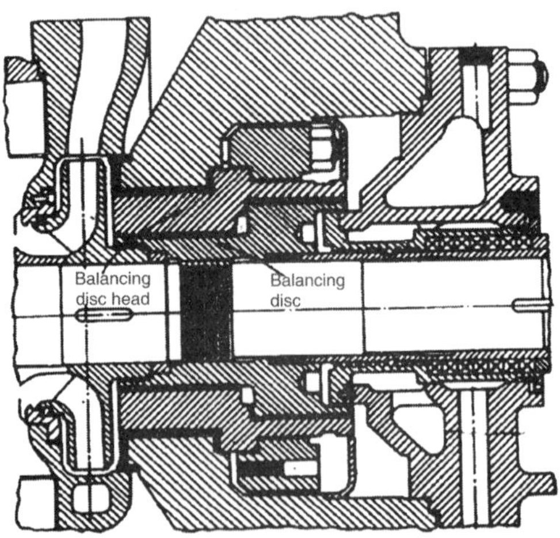

Fig. 2.73: Combination balancing disc and drum with disc located in center portion of drum (Worthington pump)

In multistage pumps, balancing disc is provided mostly for inline impeller assembly. Since the magnitude of axial thrust is very high (number of impellers multiplied by the magnitude of axial thrust for one impeller) especially in modern designs where number of stages are increased to a larger extent. Opposed impeller design system is adopted, wherein half the number of impellers will be kept facing opposite direction, in order to reduce the axial thrust. However the pump weight enormously increases due to connecting pipe between one set of impeller to another set, which is kept in opposite direction.

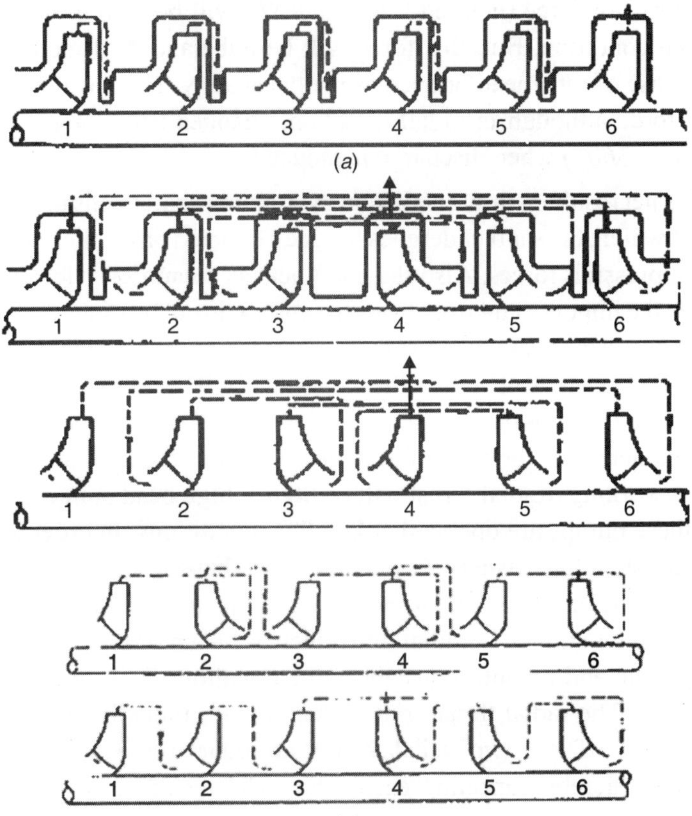

Fig. 2.74: Methods of reducing axial thrust in multi stage pump

In pumps for chemical industries with solids in suspension, (or) high viscous liquid are being pumped, balancing disc cannot be used due to the fact, solids cannot pass through the clearance between rotating and stationary members. In such cases, the axial thrust balancing is taken care of either by thrust bearing or by axial thrust balancing holes or by both.

2.8 RADIAL THRUST

At maximum efficiency conditions of pump operation, the pressure across the circumference of impeller outlet will be equal (Fig. 2.75) or near equal. At all other conditions, the pressure will be different, gradually increasing from volute tongue to volute outlet (Fig. 2.76). The direction of the pressure force will be radial and directed inwards. The resultant radial thrust or force (marked F in Fig. 2.76) will be zero or minimum at optimum condition. But, this force will be unbalanced. It increases with the increase of total head or decrease with capacity. The position of action of this force although in radial direction, changes from tongue to outlet (from 0° to 360°) when discharge is reduced.

This net resultant thrust acting on the rotating shaft bends the shaft, as a result of which, the shaft undergoes fatigue *i.e.,* acted on by alternate tension and compression forces. Also the shaft acts as a cantilever for the length between the impeller and front bearing support. Front bearings must always be designed to take care of radial thrust apart from the self weight. Although radial thrust exists at all pumps, the magnitude will be less in diffuser pumps than in volute pumps of same specification. Radial thrust depends upon the outlet diameter, and breadth of impeller as well as the total head at which pump is working. Radial thrust will be high at high head pumps especially when these pumps are operated at low flow conditions. In order to reduce radial thrust, volute casings are made into two halves, an intermediates rib provided starting from 180° to 360° *i.e.,* at the second half of the volute casing. The magnitude of radial thrust increase from 0° to 180° in first half of the volute and remains constant in the remaining 180° to 360° section (Fig. 2.77). The radial thrust in the second half gradually increases from 180° to 360°. These two radial thrusts are equal but acting opposite to each other thereby canceling each other. The resultant net radial thrust however will not be zero, in practice, but this will be taken care of by the bearings. An advantage of having double volute is that it also acts as a rib and strengthen the casing. Since radial thrust is minimum in diffuser pumps, very high head operating pumps will be always designed with diffuser followed by volute casing. Radial thrust (F_R) can be calculated from the empirical formula.

$$F_R = K(1 - (Q/Q_{opt})2.2)\,\gamma H D_2 b_2$$

$$= K.(1 - (Q/Q_{opt})2.2)\,p D_2 b_2 - \quad (2.1)$$

Where γ – Specific weight of the pumping liquid = $S\gamma_w$

S – Specific gravity of liquid γ_w – specific of water = 9.81×103 N/m^3

H – Working head (m)

p – Operating pressure (N/m^2) = $\gamma_2 H_2 = \gamma_w H_w$

D_2 – Outlet diameter of impeller (m)

b_2 – Outlet breadth of impeller (m)

K – Experimental coefficient for radial thrust

= 0.36 for n_s = 120 to 300

= 0.2 for n_s = 80 to 120

When $Q < Q_{opt}$ the direction of F_R will be acting approximately 70° from volute tongue and when, $Q > Q_{opt,}$ F_R will act approximately 270° from tongue.

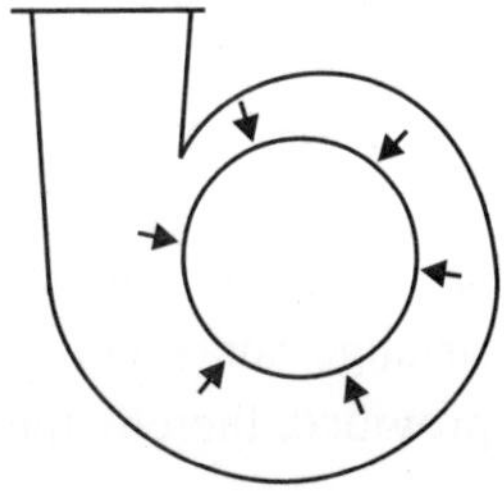

Fig. 2.75: Zero radial reaction at design capacity. Uniform casing pressure exists

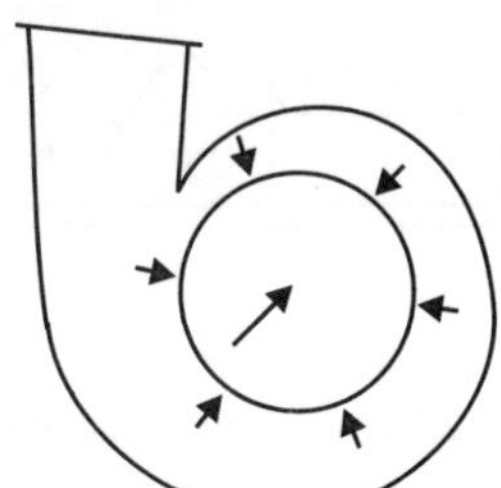

Fig. 2.76: Non uniform pressure prevails at all capacities, other than design point, above or below resulting in radial reaction *F*

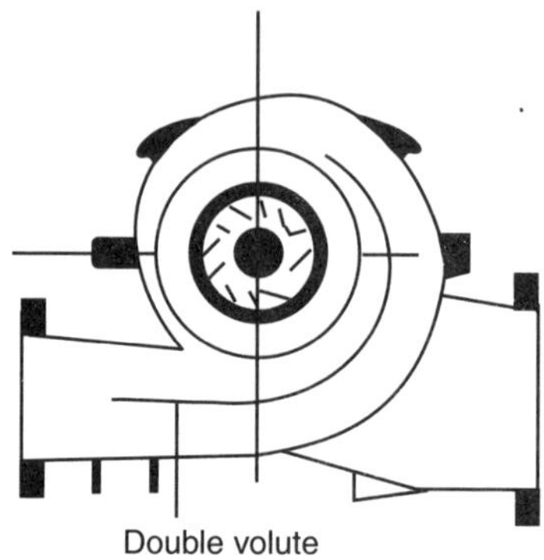

Fig. 2.77: Transverse view of a double volute casing pump

2.9 SEALING ARRANGEMENTS

2.9.1 Stuffing Box Arrangements

Constructional features of different forms of stuffing boxes with lantern ring and flushing arrangement are given in Figs. 2.80–2.87. Each arrangement is to be adopted according to the functional aspects of the pump such as temperature, pumping solution etc. It is the usual practice to keep the stuffing box pressure at the packing slightly above atmospheric pressure (1.5 to 1.75 kg/cm^2 above atmosphere). This reduces the leakage through the gland as well as long life for packing. In case, where the pressure is higher, flushing arrangements with lantern rings or with proper wearing ring clearance are provided, thereby the pressure is brought down to the standard value.

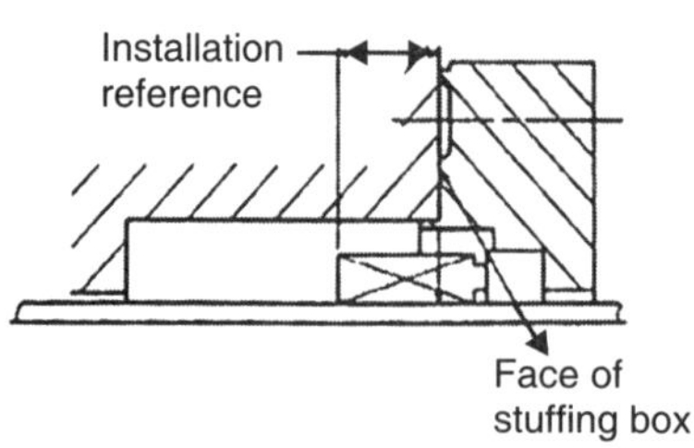

Fig. 2.78: Typical installation

Fig. 2.79: Lantern ring (also called as seal gagues)

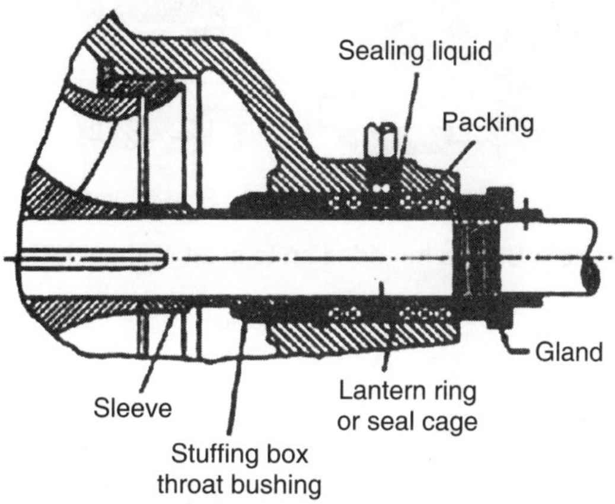

Fig. 2.80: Conventional stuffing box with threaded bush

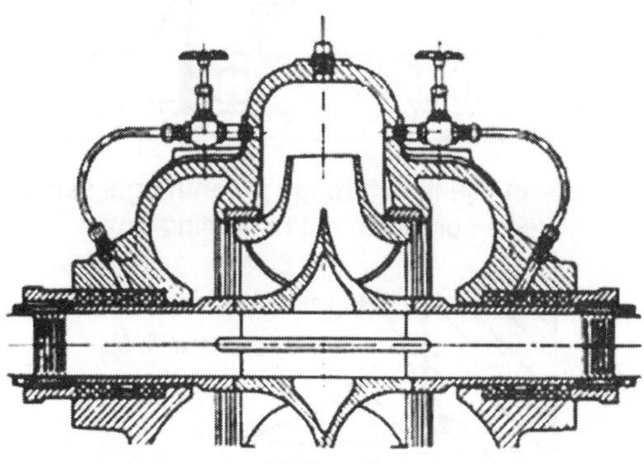

Fig. 2.81: Piping connections from discharge to lantern ring

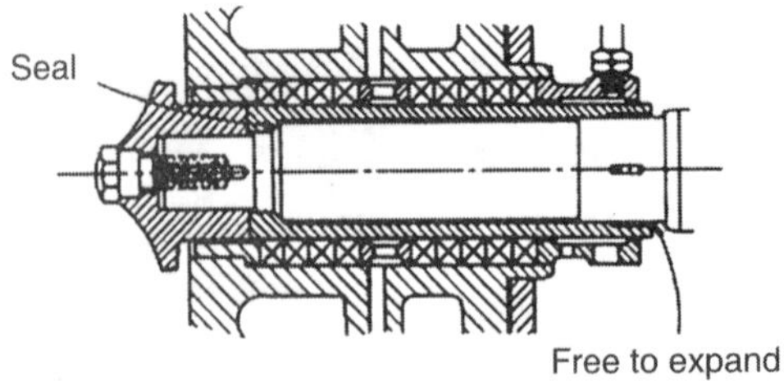

Fig. 2.82: Sleeve with 45° level contacting surface

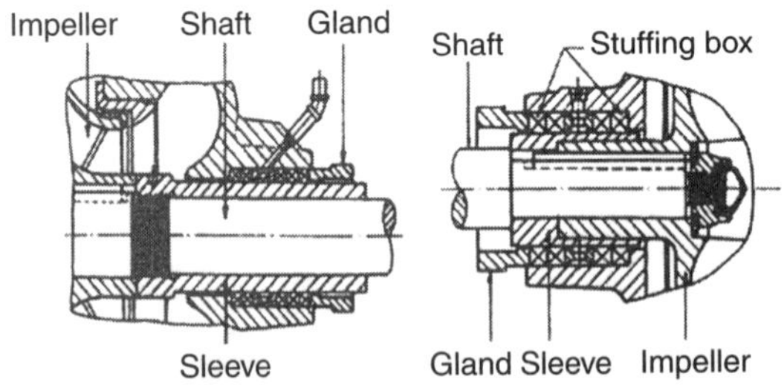

Fig. 2.83: Sleeve threaded on to shaft, without lock nut

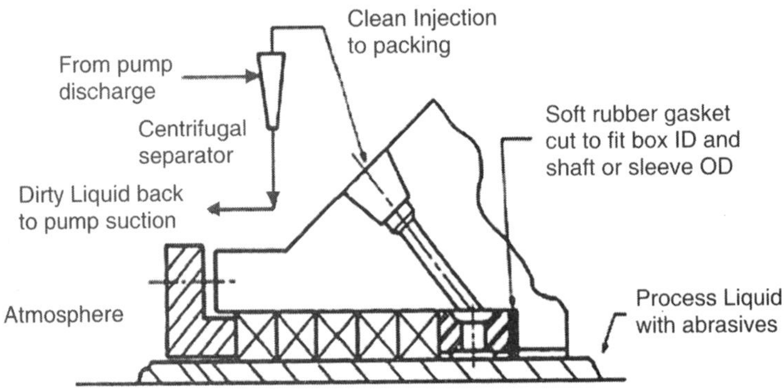

Fig. 2.84: Clean injection through a centrifugal separator to keep abrasives out of stuffing box

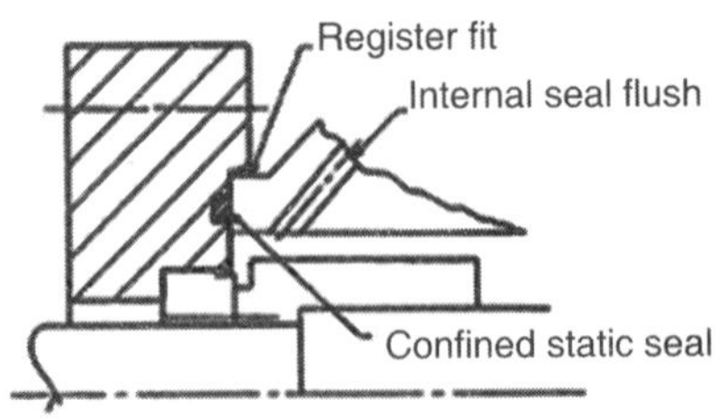

(a)

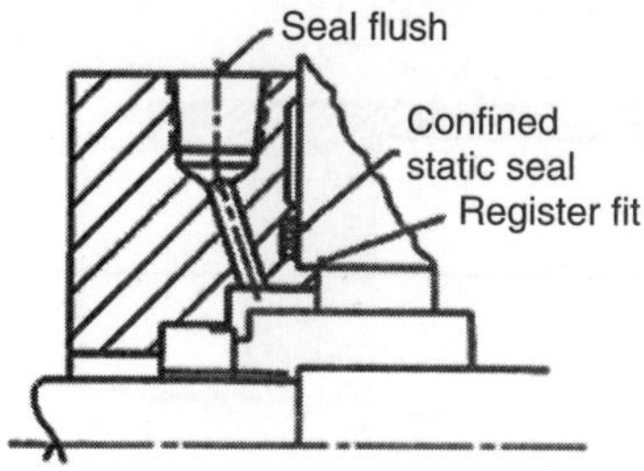

(b)

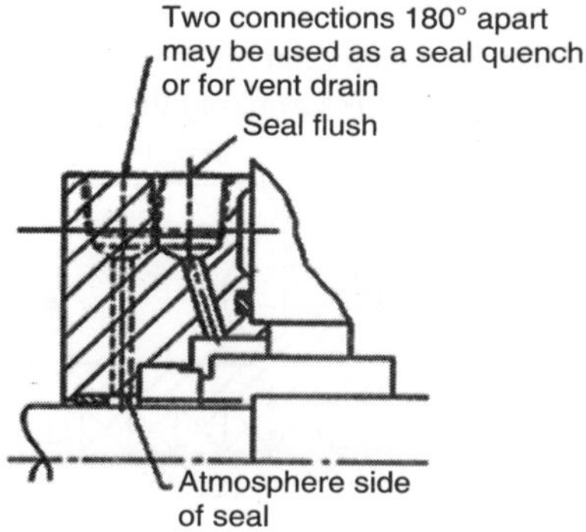

(c)

Fig. 2.85: Basic gland plate designs: (a) Plain gland plate (b) Flush gland plate (c) Quench or vent and drain gland plate

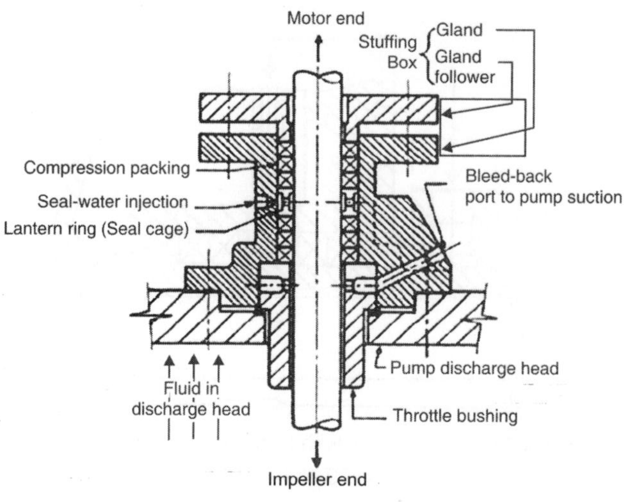

Fig. 2.86: Compression packing with throttle bushing for pressure breakdown

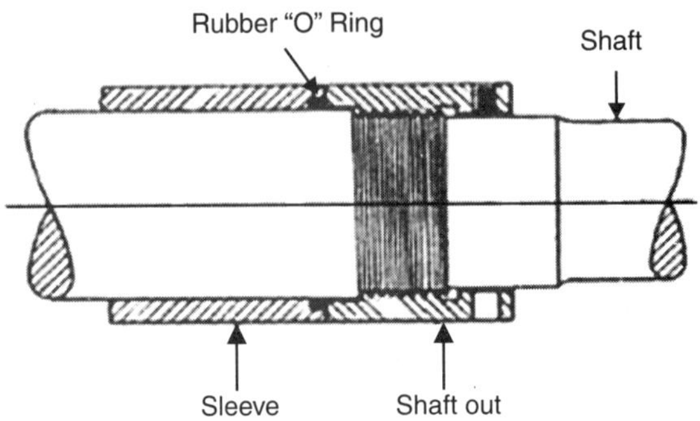

Fig. 2.87: Seal arrangement for shaft sleeve to prevent leakage along the shaft

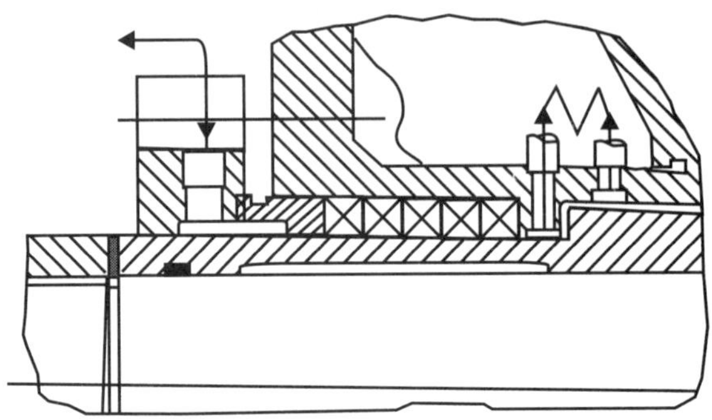

Fig. 2.88: Stuffing box with classic cooling water jacket cooling the outer diameter of the gland

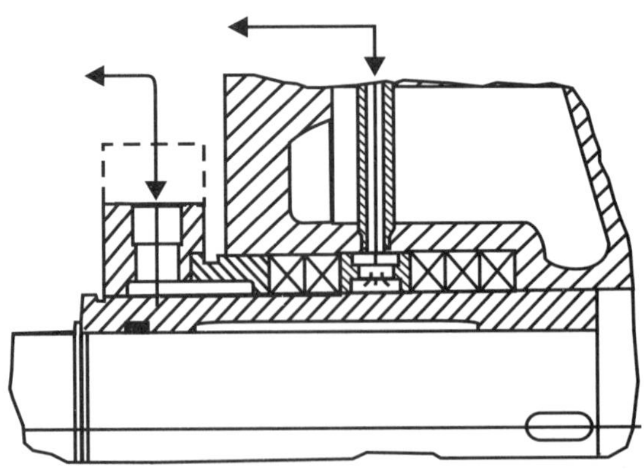

Fig. 2.89: Stuffing box with intermediate lantern ring for sealing water supply

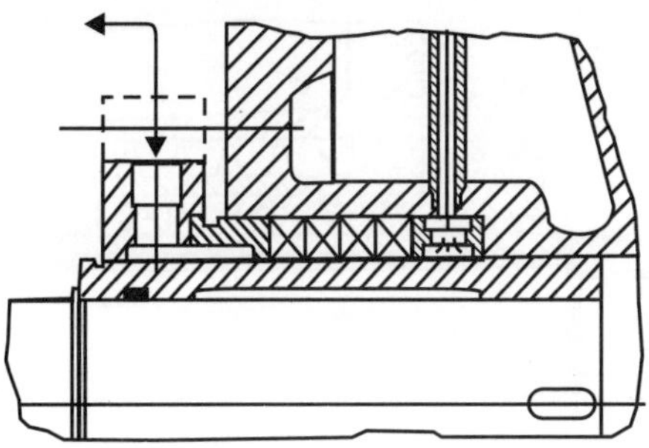

Fig. 2.90: Stuffing box with lantern type enduring ring for cold water injection

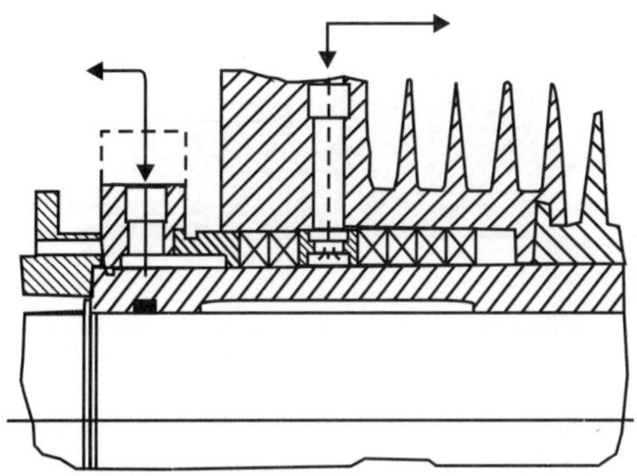

Fig. 2.91: Stuffing box of special design with hollow shaft sleeve to cool the inner diameter of the gland

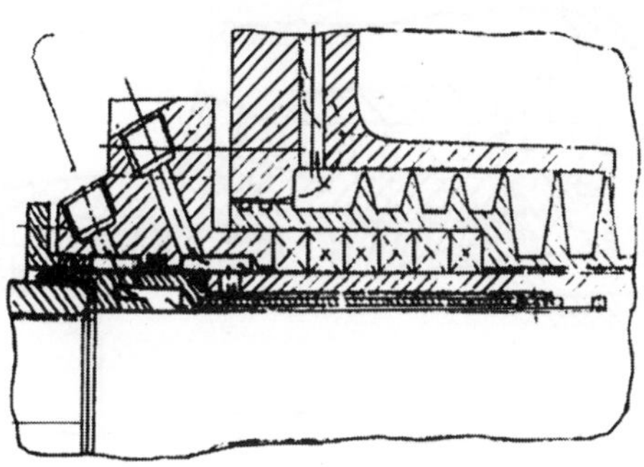

Fig. 2.92: Stuffing box with double cooling effect and duplicate cooling feed, cooling inner and outer diameter of the gland

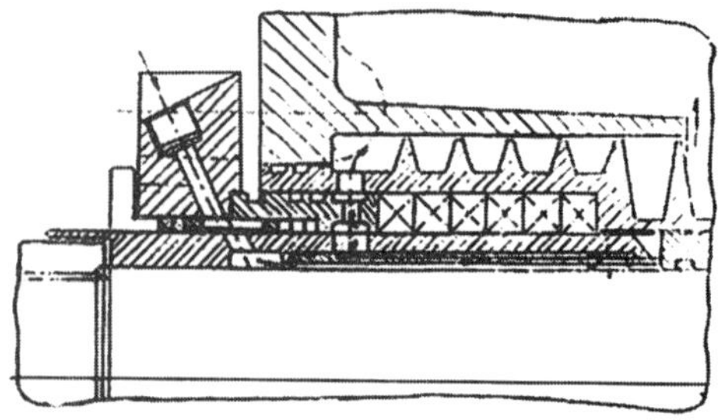

Fig. 2.93: Stuffing with double cooling effect and single cooling feed, cooling inner and outer diameter of gland

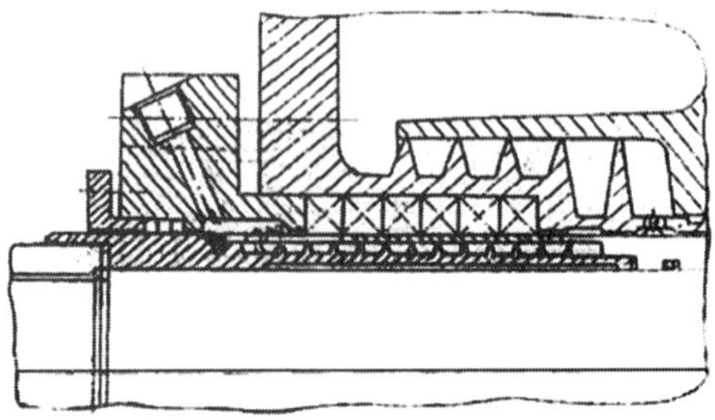

Fig. 2.94: Stuffing box, with double cooling effect, with single cooling feed and also introducing cooling liquid upstream of the packing end ring, for cooling inner and outer diameter of gland

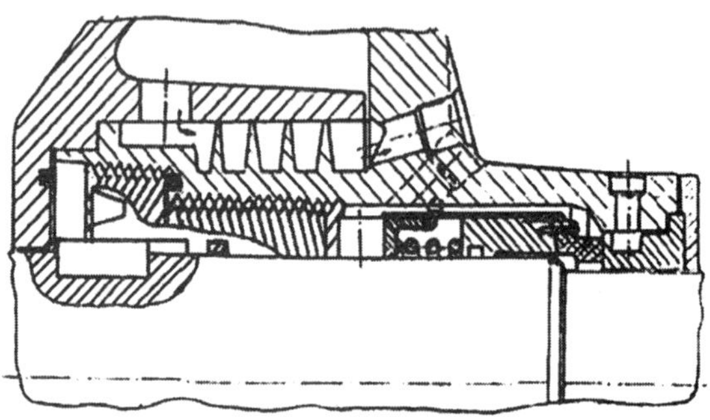

Fig. 2.95: Gland area of feed pump with injection type shaft seal, intensive cooling and differential type balancing device absorbing pressure fluctuations of feed pump suction pressure, sealing water pressure

2.9.2 Procedure for Installation of Continuous Coil Packing

1. Loosen and remove the gland from the stuffing box.
2. Using a packing puller, begin to remove the old packing rings.
3. Remove the split lantern ring and then continue removing packing with the puller.
4. After the packing has been removed, check the sleeve for scoring and nicks. If the shaft sleeve or shaft cannot be cleaned up, it must be replaced. Check the size of the stuffing box bore and the shaft sleeve or shaft diameter to determine what size packing should be used.
5. After the size of the packing has been determined, wrap the packing tightly around a mandrel, which should be the same size as the pump shaft or sleeve. The number of coils should be sufficient to fill the stuffing box. Cut the packing along one side to form individual rings.
6. Before beginning the assembly of any packing material, be sure to read all instructions from the manufacturer. Assemble the split packing rings on the pump. Each ring should be seated individually, with the split ends staggered 90° and the gland tightened to seat and fully compress the ring. Be sure the lantern ring is reinstalled correctly at the flush connection. Then pack off the gland and retighten only finger-tight. The exception to this procedure is that TFE packing should be installed one ring at a time but not seated because TFE packing have high thermal expansion.
7. Allow excess leakage during break in, to avoid the possibility of rapid expansion of the packing, which could score the shaft sleeve or shaft so that leakage could not be controlled.
8. Leakage should be generous upon start-up. If the packing begins to overheat at start-up, stop the pump and loosen the packing until leakage is obtained. Restart only if the packing is leaking.

Continuous lubrication should be introduced at the lantern ring. Table is only a guide. Consult packing manufacturer with complete operating conditions for exact recommendation. Pressure relates to the operating pressure at the stuffing box. PV data given is based upon a 2 in (5.08 cm) shaft at 1750 and 3600 rpm. Temperature means the product temperature.

Table 2.7: Service limitations of common packing materials

Packing material	Pressure (max) lb/in² gauge (kPa gauge)	PV rating (max) c, lb/in² gauges.fpm (bar.m/s)	Temp. (max) d. °F (°C)	PH range	Comments
Cotton	100 (689)	188.000 (65.8)	150 (65.6)	5–7	Nonabrasive material, for cold water and dilute salt solutions
Flax/ramine	100 (689)	188.000 (65.8)	150 (65.6)	5–7	High wet strength and excellent resistance to fungi and rotting, for cold water and dilute salt solutions
Plastic	100 (689)	188.000 (65.8)	600 (315.5)	4–8	Excellent sealing qualities: reacts well to gland adjustments; can extrude at higher pressures if not backed up by brainded or asbestos packing
	250 (1723)	471.000 (165)	150 (65.6)		
Asbestos, grease or oil impregnated	100 (689)	188.000 (65.8)	750 (398.8)	4–8	For hot or cold water, brine, oil, mild caustics, solvents and acids
	250 (1723)	471.000 (165)	500 (260)		
Asbestos, TFE-impregnated	250 (1723)	471.000 (165)	500 (260)	2–10	For mild chemicals and solvents
Lead	250 (1723)	471.000 (165)	450 (232.2)	2–10	Shaft sleeve must have Brinell hardness of 500 or more; for hot oils and boiler-feed water
Aluminum or Copper	250 (1723)	471.000 (165)	750 (398.8)	3–10	Shaft sleeve must have Brinell hardness of 500 or more; for hot oils and boiler-feed water
TFE filament	250 (1723)	471.000 (165)	500 (260)	0–14	For corrosive liquids and food service: usually requires slightly higher break in leakage
Aramid fiber	250 (1723)	471.000 (165)	500 (260)	3–10	Strong reslient packing; maximum speed 1900 fpm (9.6 m/s); good in abrasives and chemicals

Graphite/ carbon filament	250 (1723)	471.000 (165)	750 (398.8)	0–14	For corrosive liquids and high temperature applications
Gra foil	250 (1723)	471.000 (165)	750 (398.8)	0–14	Excellent conductor of heat from the sealing surfaces; operates with minimum leakage; excellent radiation resistance

Table 2.8: Packing troubles, causes and remedies

Trouble	Cause	Remedies
No liquid delivered by pump	Lack of prime (packing loose or defective, allowing air to leak into suction)	Tighten or replace packing and prime pump
Not enough liquid delivered by pump	Air leaking into stuffing box	Check for leakage through stuffing box while operating- if no leakage after reasonable gland adjustment, new packing may be needed. or Lantern ring may be clogged or displaced and may need centering in line with sealing liquid connection or Sealing liquid line may be clogged. or Shaft or shaft sleeve below packing may be badly scored, allowing air to be sucked into pump.
	Defective packing	Replace packing and check smoothness of shaft or shaft sleeve.
Not enough pump pressure	Defective packing	As for proceeding
Pump works for a while and quits	Air leaks into stuffing box	As for proceeding

Pump takes too much power	Packing too tight	Release gland pressure and retighten reasonably.
		Keep leakage flowing if none, check packing, sleeve, or shaft.
Pump leaks excessively at stuffing	Defective packing	Replace worn packing. Replace packing damaged by lack of lubrication.
	Wrong type of packing	Replace packing not properly installed or run in. Replace improper packing with correct grade for liquid being handled.
	Shaft or shaft sleeve scored	Put in lathe and machine-true and smooth or replace.
Stuffing box overheats	Packing too tight	Release gland pressure
	Packing not lubricated	Release gland pressure and replace all packing if any burnt or damaged.
	Wrong grade of packing	Check with pump of packing manufacturer for correct grade.
	Insufficient cooling water to jackets	Check for open supply line valves or clogged line.
	Stuffing box improperly packed	Repack
Packing wears too fast	Shaft or shaft sleeve worn or scored	Remachine or replace
	Insufficient or no lubrication	Repack, making sure packing loose enough to allow some leakage.
	Packing packed improperly	Repack properly, making sure all old packing removed and box clean.
	Wrong grade of packing	Check with pump or packing manufacturer.
	Pulsating pressure on external seal liquid line	Remove cause of pulsation.

2.9.3 CAUSES FOR SHORT PACKING LIFE

In order that the packing should be operated properly, the equipment must be in good condition. Shafts should be checked for run out and eccentricity to ensure that they are within the manufacturer's recommended tolerances. Surfaces in contact with the packing should be finished to the correct smoothness and tolerance. Table 2.8 lists common troubles that affect packing life, causes and possible cures are also given.

2.10 MECHANICAL SEAL

Mechanical seals are commonly used in pumps, as replacement for packing materials. In some of the occasions, only mechanical seal can be used. Example: Pumping of fuming liquids, corrosive liquids etc. Mechanical seals are available in different forms for different pumping solutions at different operating pressures and temperatures. It is always better to consult seal manufactures for proper selection of seal.

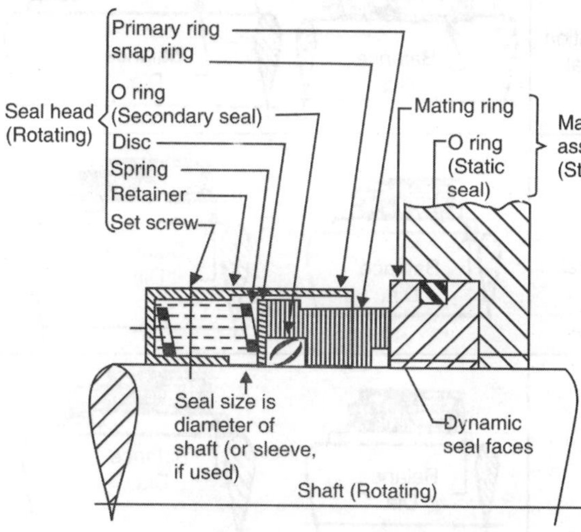

Fig. 2.96: Mechanical seal

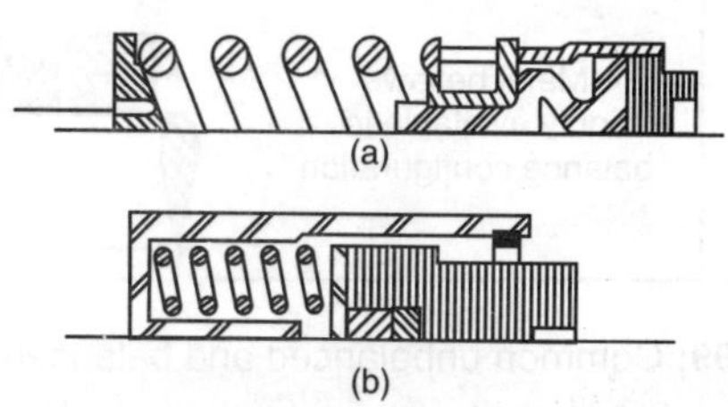

Fig. 2.97: Comparison of (a) Single-spring and (b) Multiple-spring seals

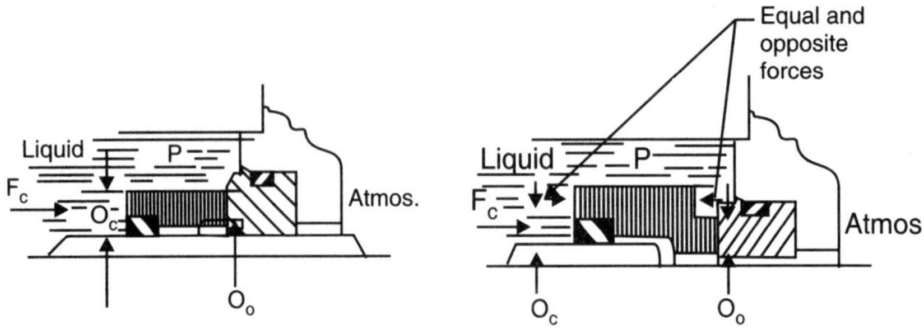

(a) (b)

Fig. 2.98: Hydraulic-pressure acting on the primary ring: (a) balanced (b) unbalanced

Type of seal \ Type of balance	Unbalanced b = 100% or >100%	Balanced b < 100%
Full convolution bellow seal	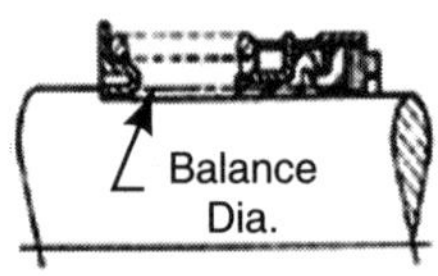	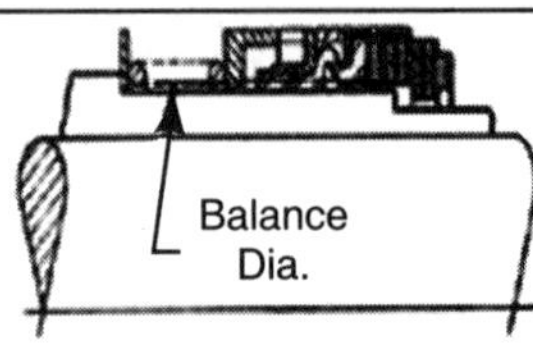
Wedge seal	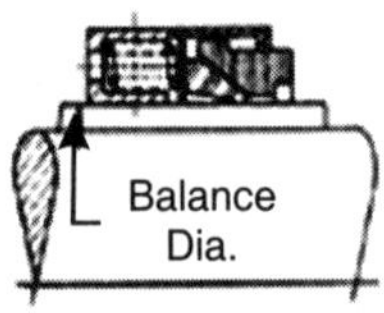	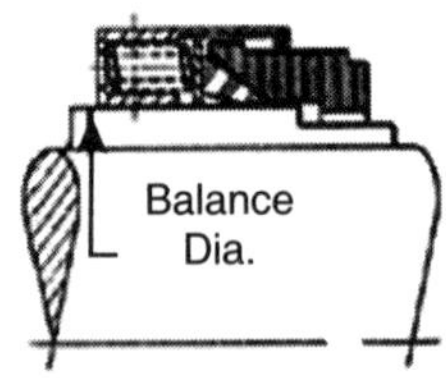
O ring seal	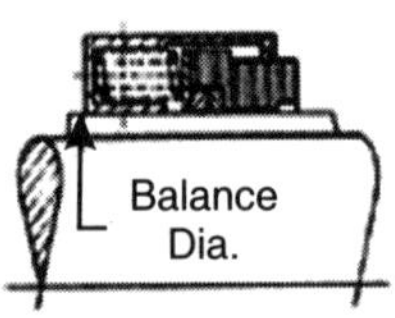	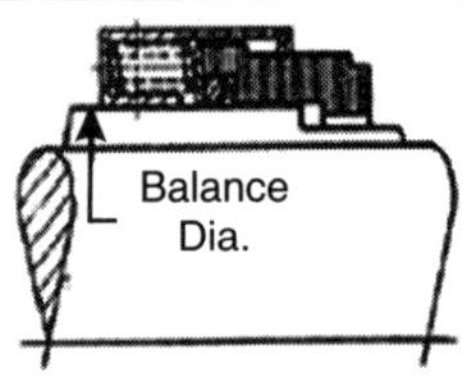
Multiple convolution Metal bellows seal	Metal bellows only available in balance configuration	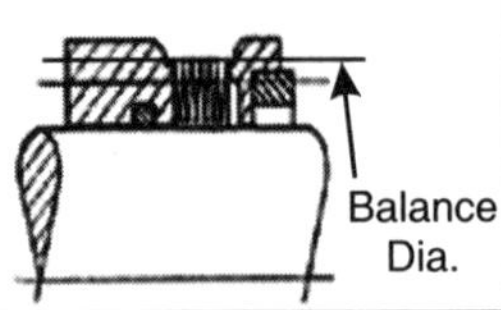

Fig. 2.99: Common unbalanced and balanced seal

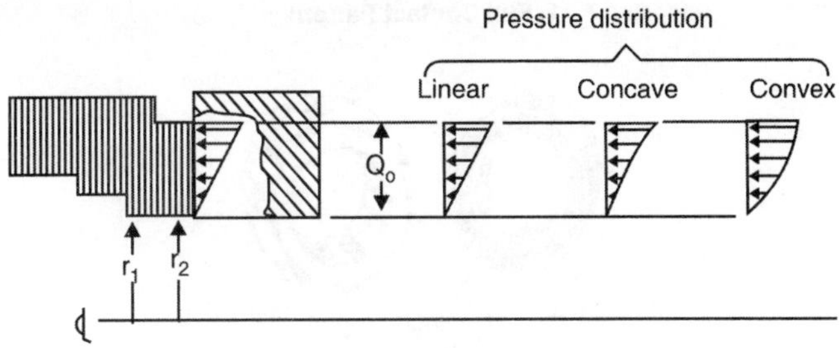

Fig. 2.100: Pressure wedge

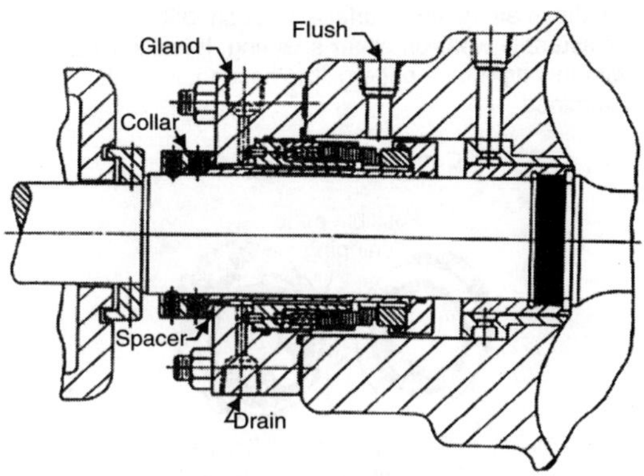

Fig. 2.101: Stationary seal with rotating mating ring (John Crane–Houdaille, Inc.)

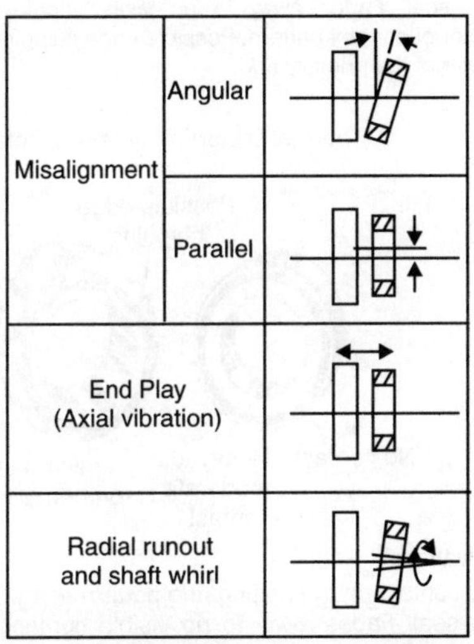

Fig. 2.102: Common types of motion that influence the seal performance

1. Full Contact Pattern

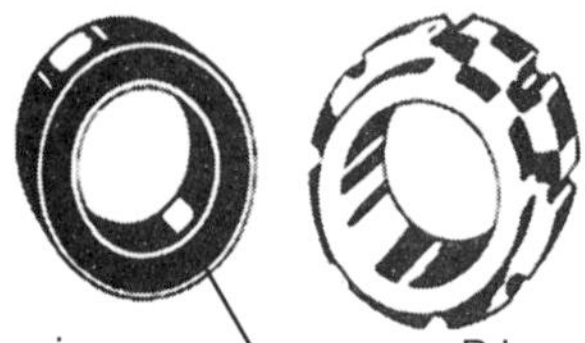

Observation:

Typical contact pattern for a non-leaking seal. Full contact on the matting ring surface through 360°. Little or no measurable wear on either seal ring. If leaking is present with this type face pattern the secondary seals must be examined.

2. Coning (negative rotation)

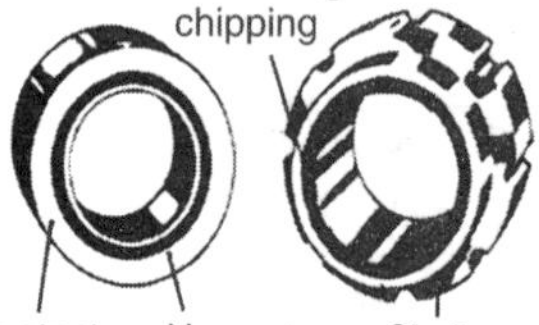

Observation:

Heavy contact on the mating ring pattern at the outside diameter of the seal. Fades away to no visible contact at the inside diameter of contact pattern. Possible edge chipping on the outside diameter of the primary ring.

3. Thermal Distortion (Positive Rotation)

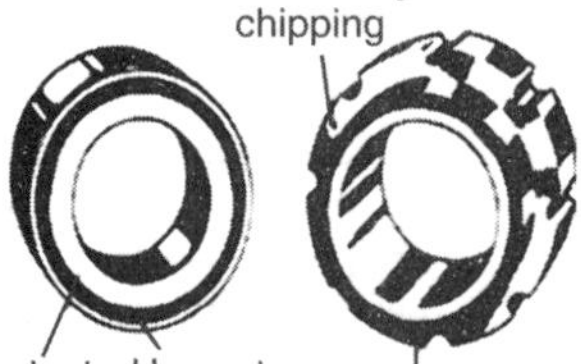

Observation:

Heavy contact on the mating ring pattern at the inside diameter of the seal. Fades away to no visible contact at the outside diameter of contact pattern. Possible edge chipping on the inside diameter of the primary ring.

4. Mechanical Distortion

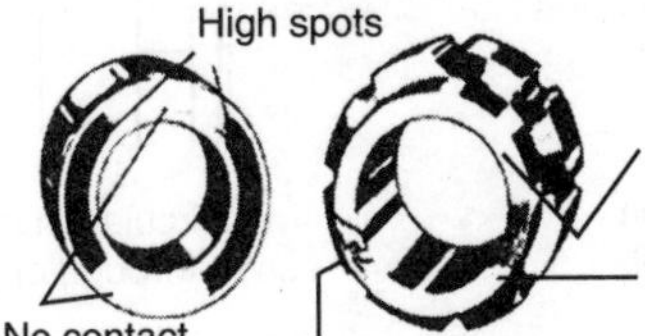

Observation:

Mating ring is distorted mechanically, creating two large contact spots. Pattern fades away between contact areas.

5. Mechanical Distortion

Observation:

Mating ring is being distorted mechanically, creating contact through approximately 270°. Pattern faces away at the low spot

6. Mechanical Distortion

Observation:

Mating ring is being distorted mechanically, creating contact at bolts. High spots are at each bolt location

Fig. 2.103. Causes of seal leakage (John Crane–Houdaille, inc.)

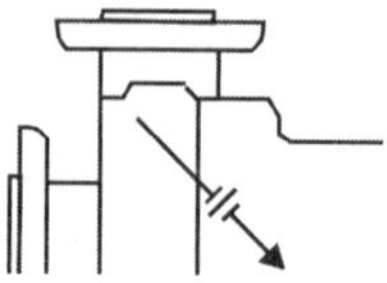

Integral (internal) recirculation from pump discharge to seal

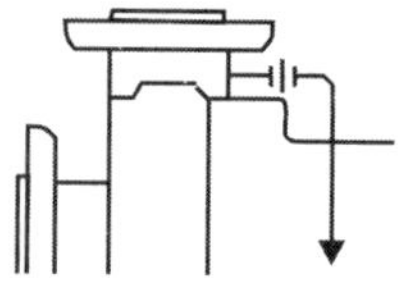

Recirculation from pump case through orifice to seal

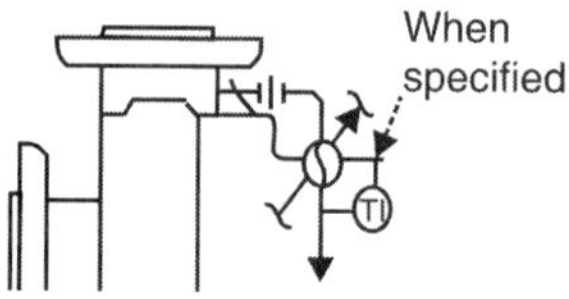

Recirculation from pump case through strainer, orifice and cooler to seal

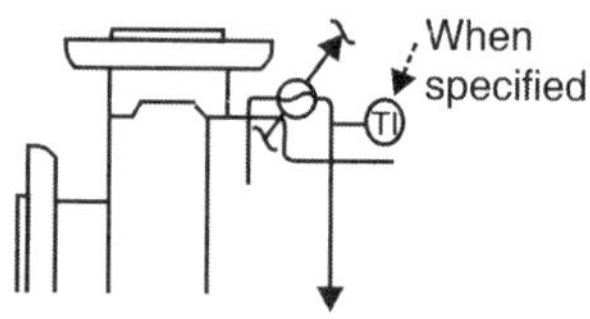

Recirculation from seal with pumping ring through cooler and back to seal

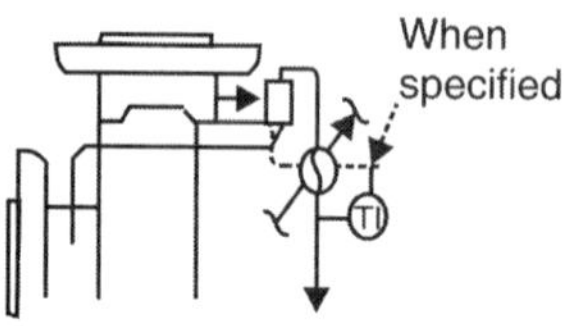

Recirculation from pump case through cyclone separator delivering clean fluid through cooler to seal and fluid with solids back to pump suction

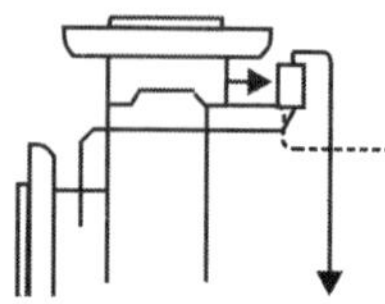

Recirculation from pump case through cyclone separator delivering clean fluid through cooler to seal and fluid through cooler to seal and fluid with solids back to pump suction

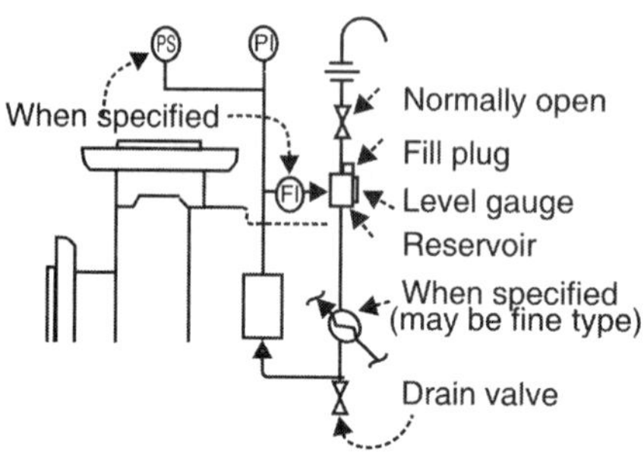

External fluid reservoir (see Note 2). non pressurized: Thermosyphon or forced circulation as required

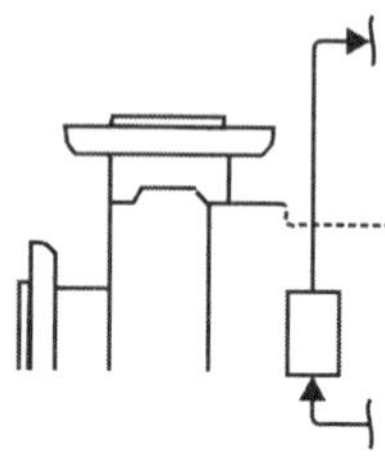

Circulation of clean fluid from an external system (see Note 2).

Legend

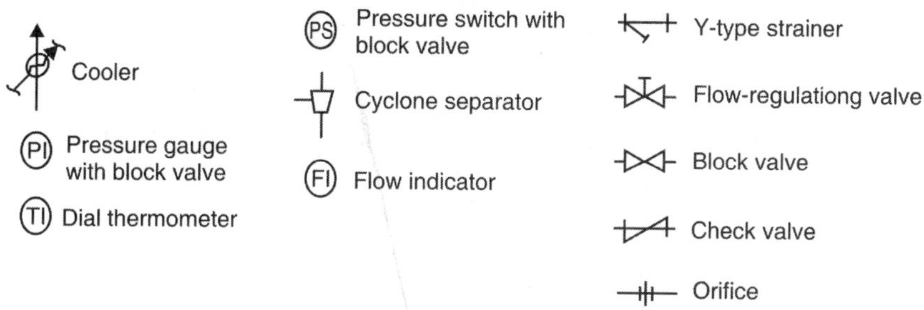

Notes:

1. These plans represent commonly used systems. Other variations and systems are available and should be specified in details by the purchaser or mutually agreed upon by the purchaser and the vendor.
2. For plan 32, the purchaser shall specify the fluid characteristics, and the vendor shall specify the volume (gallons per minute) and pressure (pounds per square inch gauge) required.

Fig. 2.104: Piping for mechanical seals, from API Standards 610, 6th ed. (Reproduced from American Petroleum Institute)

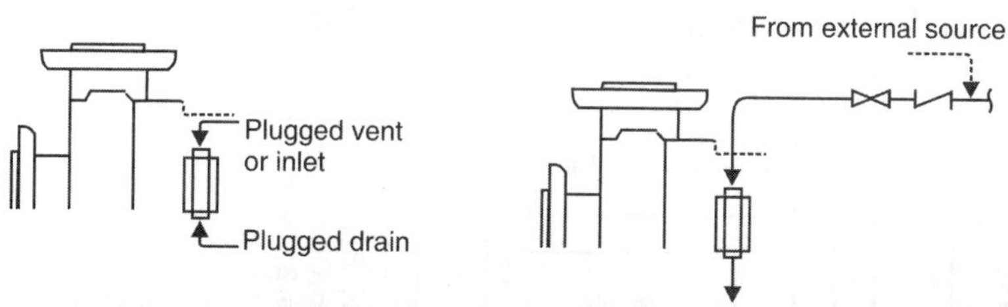

Tapped connections for purchaser's use: Note 2 shall apply when purchaser is to supply fluid (steam, ges, water, etc.) to auxiliary sealing device.

External fluid quench (steam, gas, water, etc. see Note 2).

Notes:

1. These plans represent commonly used systems. Other variations and systems are available and should be specified in details by the purchaser mutually agreed upon by the purchaser and the vendor.
2. The purchaser shall specify the fluid characteristics, when supplemental seal fluid is provided. The vendor shall specify the volume (gallons per minute) and pressure (pounds per square inch gauge) required. When these are factors.
3. See Figure D-2 for explanation of symbols not specified here.

Fig. 2.105: Piping for throttle bushing, auxiliary seal device, tandem seals, or double seal from API standard 610, 6th ed. (Reproduced by courtesy of the American Petroleum Institute)

Table 2.9: Temperature limits of secondary seal materials

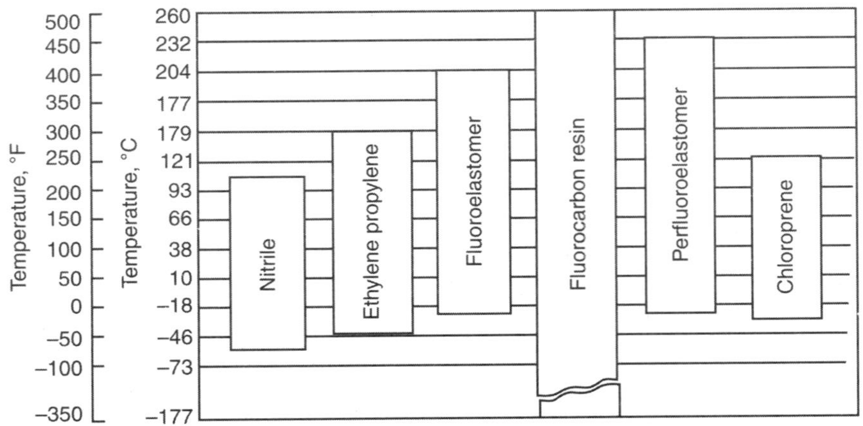

A piping arrangement , or plan, defines how a seal installation will be cooled or, in some cases, heated. Commonly used systems have been defined by API and are shown in Figs. 2.104 and 2.105.

Table 2.10: Properties of common seal materials

Property	Cast iron	Ni-resist	Ceramic		Carbides		Caron			
			85% (Al_2O_3)	99% (Al_2O_3)	Tungsten 6%(O)	Silicon Si-C	Resin	Babbitt	Bronze	Si-C conv.
Modulus of elesticity × 10^6 lb/in^2(× 10^3 Mpa)	13-15.95 (90-110)	10.5–16.9 (72–117)	32 (221)	50 (245)	90 (621)	48–57 (331–393)	2.5–4.0 (17.2–27.6)	1.04–4.1 (7.2–28.3)	2.9 –4.4 (20–30)	2–2.3 (13.8–15.9)
Tensile strength × 10^6 lb/in^2(× 10^3 Mpa)	65-120 (448-827)	20–45 (138–310)	20 (138)	39 (269)	123.25 (850)	20.65 (142)	4.5–9 (31–62)	8–8.6 (55–59)	7.5–9 (52–62)	2 (14)
Coefficient of thermal expansion × 10^{-6} lb/in^2, F(cm/cm.K)	6.6 (11.88)	6.5–6.8 (11.7–12.24)	3.9 (7.02)	4.3 (7.74)	2.53 (4.55)	1.88 (3.38)	2.3–3.4 (4.14–6.12)	2.1–2.7 (3.78–14.86)	2.4–3.1 (4.32–5.58)	2.4–3.2 (4.32–5.76)
Thermal conductivity Btu.ft/h. ft^2.F	23-29 (39.79-50.17)	25–28 (43.25–48.44)	8.5 (14.70)	14.5 (25.08)	41–48 (70.93 – 83.04)	41–60 (70.93 – 10.38	3.8–12 (6.57–20.76	6–9 (10.38–15.57)	8–8.5 (13.84–14.70)	30 (51.9)

Density lb/ in^3(kg/m^3)	0.259-0.268 (7169-7418)	0.264–0.268 (7307-7418)	0.123 (3405)	0.137 (3792)	0.59 (6,331)	0.104 (2879)	0.064 –0.069 (1771–1910)	0.083–0.112 (2297–3100)	0.083–0.097 (2297–2685)	0.067–0.070 (1854–1938)
Hardness	Brinell		Rockwell A			Rockwell 45 N	Shore			Rockwell 15T
	217–269	131–183	87	87	92	86–88	80–105	60–95	70–92	90

Table 2.11: Frequently used seal face materials and their PV limitations

Sliding material		PV limit lb/n². ft/ min (bar.m/s)	Comments
Rotating	Stationary		
Carbon–Graphite	Ni-resist	100,000 (35.03)	Better thermal shock resistance than ceramic
	Ceramic (85% Al_2O_3)	100,000 (35.03)	Poor thermal shock resistance and much better corrosion resistance than Ni-resist
	Ceramic (99% Al_2O_3)	100,000 (35.03)	Better corrosion resistance than 85% Al_2O_3 ceramic
	Tungsten carbide (6% CO)	500,000 (175.15)	With bronze-filled carbon graphite, PV is upto 100,00 lb/in². ft/min (35.03bar.m/s)
	Tungsten carbide (6% CO)	500,000 (175.15)	Ni Binder for better corrosion resistance.
	Silicon carbide (solid)	500,000 (175.15)	Better corrosion resistance than tungsten carbide but poorer thermal shock resistance.
	Silicon carbide converted carbon	500,000 (175.15)	Good wear resistance thin layer of si.c make refix-aping questionable
Carbon–Graphite		500,000 (175.51)	Low PV, but very good against face blistering
Tungsten carbide		120,000 (42.04)	PV is up to 185,000 lb/ in².ft/min (64.8 bar.m/s) with two grades that have different % of binder.
Silicon carbide converted carbon		500,000 (175.15)	Excellent abrasion resistance; more economical than solid silicon carbide.
Silicon carbide solid		500,000 (175.15)	Excellent abrasion resistance, good corrosion resistance, and moderate thermal shock resistance.
Ceramic		10,000 (3.50)	Good service on sealing paint pigments

Table 2.12: Causes of seal leakage

Symptom	Possible causes	Corrosive procedures
Seal spits and sputters ("face popping") in operation	Seal fluid vaporizing at seal interfaces	Increase cooling of seal faces check for proper seal balance with seal manufacturer. Add bypass flush line if not in use. Enlarge bypass flush line and/or orifices in gland plate check for seal interface cooling with seal manufacturer.
Seal drips steadily	Faces not flat carbon graphite seal faces blistered seal faces thermally distorted	Check for incorrect installation dimensions Check for improper materials or seals for the application improve cooling flush lines Check for gland plate distortion due to over torquing of gland bolts. Check gland gasket for proper compression. Clean out foreign particles between seal faces; relap faces if necessary. Check for cracks and chops at seal faces; replace primary and mating rings.
	Secondary seals nicked or scratched during installation O rings overaged. Secondary seals hard and brittle from compression set. Secondary seals soft and sticky from chemical attack.	Replace secondary seals Check for proper lead in chamfers, burrs etc., Check for proper seals with seal manufacturer Check with seal manufacturer for other material.
	Spring failure hardware damaged by erosion drive mechanisms corroded	Replace parts Check with seal manufacturer for other material
Seal squeals during operation	Amount of liquid inadequate to lubricate seal faces	Add bypass flush line if not in use. Enlarge bypass flush line and/or orifices in gland plate

Carbon dust accumulates on outside of gland ring	Amount of liquid inadequate to lubricate seal faces. Liquid film evaporating between seal faces.	Add bypass flush line if not in use enlarge bypass flush line and/or orifices in gland plate. Check for proper seal design with seal manufacturer if pressure in stuffing box is excessively high.
Seal leaks	Nothing appears to be wrong	Refer in list under "Seal drips steadily" Check for squareness of stuffing box to shaft. Align shaft, impeller, bearing, etc., to prevent shaft vibration and/ or distortion of gland plate and/or mating ring.
Seal lift is short	Abrasive fluid	Prevent abrasives from accumulating at seal faces. Add bypass flush line if not in use. Use abrasive separator of filter
	Seal running too hot	Increasing cooling of seal faces. Increase bypass flush line flow. Check for obstructed flow in cooling lines.
	Equipment mechanically out of line	Align Check for rubbing of seal on shaft.

The following Tables 2.13 and 2.14 give the allowance and tolerance used for fitting mechanical seals into the pumps shaft.

Table 2.13: Press fit allowance and tolerance in the out of roundness and outside diameter of shaft seals

Outside diameter (mm)	Press fit allowance		Permissible quality of roundness	
	Rubber OD (mm)	Metal OD (mm)	Rubber OD (mm)	Metal OD (mm)
Upto 50	+ 0.15 + 0.30	+ 0.05 + 0.13	0.20	0.05
Above 50 to 80	+ 0.2 + 0.3	+ 0.1 + 0.2	0.25	0.10

Above 80 to 120	+ 0.2 + 0.35	+ 0.12 + 0.22	0.30	0.15
Above 120 to 180	+ 0.25 + 0.40	+ 0.15 + 0.25	0.30	0.20
Above 180 to 300	+ 0.25 + 0.40	+ 0.18 + 0.28	0.35	0.20
Above 300 to 500	+ 0.30 + 0.50	+ 0.18 + 0.30	0.40	0.25

For seal ID (Vs) shaft diameter		
Shaft Diameter	**Seal ID less by mm**	**Tolerance**
25	Less by 1.5	± 0.2
50	Less by 2.0	± 0.2
100	Less by 2.5	± 0.3
200	Less by 3.0	± 0.3
300	Less by 3.5	± 0.4
400	Less by 4.0	± 0.5
500	Less by 4.0	± 0.5

Table 2.14: General tolerances for machining operations of components for dimension, which are not mentioned with specific tolerances

(A) Lengths	
Dimensions	**Tolerance (mm)**
(a) Over 10 upto 4 (b) Over 4 upto 16 (c) Over 16 upto 63 (d) Over 63 upto 250 (e) Over 250 upto 1000 (f) Over 1000	± 0.10 ± 0.20 ± 0.30 ± 0.50 ± 0.80 ± 1.20
(B) Diameters	
(a) Over 1 upto 4	0.00 0.10
(b) Over 4 upto 16	0.00 0.20

(c) Over 16 upto 63 (d) Over 63	0.00 0.30 0.00 0.50
(C) Holes	
(a) Over 1 upto 4 (b) Over 4 upto 8	0.00 + 0.20 0.00 0.30 + 0.20
(c) Over 8 upto 16 (d) Over 16 upto 32 (e) Over 32 upto 63 (f) Over 63	0.00 + 0.40 0.00 to + 0.50 0.00 to + 0.60 0.00 to + 0.80
(D) Depth of drill and tapped holes	
(a) Over 4 upto 16 (b) Over 16 upto 63 (c) Over 63	± 1.00 ± 0.50 ± 2.00
(E) Deviation of hole centres	
Holes Size (mm) (a) Upto 4 (b) Over 4 upto 16 (c) Over 16 upto 63 (d) Over 63	Deviation from Drawing ± 0.3 ± 0.5 ± 0.7 ± 1.2

Note: For holes on 'PCDs' the tolerance is on the chord dimension and on the distance from the Spigot.

3 Testing and Performance Analysis

All roto dynamic and positive displacement type pumps, are tested normally for load test to determine total head–discharge (H–Q), power-discharge (N–Q) and efficiency-discharge (η–Q) characteristics, for different capacities, from full close to full open of the control valve. Cavitation test is conducted to determine NPSH values of the pump at different working regions.

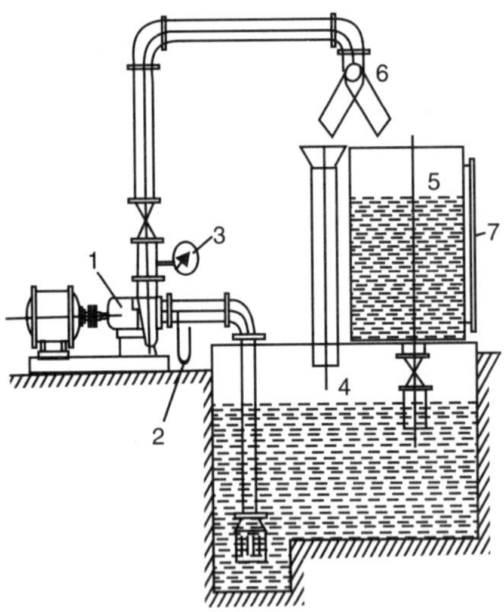

1. Pump
2. Suction vacuum manometer
3. Delivery pressure gauge
4. Auxiliary pipe
5. Delivery tank
6. Water diverting chute
7. Scale for height measurement

Fig. 3.1: Load test stand for pump

3.1 LOAD TEST

Figure 3.1 shows the line sketch of a test rig for pump testing. Essentially, it consists of a prime mover, suction sump or tank, delivery tank, suction and delivery pipes, gate valve to regulate the flow and measuring arrangements for flow rate, total head, power and speed. This test set up is normally called 'Load Test Rig'.

The pump 1, to be tested, is mounted on a base, with the prime mover. The suction and delivery pipes are connected to the pump with sufficient straight length near the pump before suction eye and after volute outlet, to ensure uniform flow condition. For accurate testing, the straight lengths at suction and delivery must be at least 4 to 6 times the diameter of the corresponding pipe.

A gate valve is provided at the delivery to regulate the flow. Flow rate is measured by volumetric method *i.e.*, by the collecting tank method. The time taken for collecting the pumping fluid for a specific rise in the tank 5 is determined from the scale 7 fitted at the side of the tank. The flow rate is determined from the known dimensions of the collecting tank 5 and 7, the rise in the fluid level, in the tank 7, and the time taken for that rise. For smaller flow rates, a venturi meter or orifice meter, and for larger flow rates, open channel flow, with V-notch are used. A foot valve and strainer at the bottom of the suction pipe ensures the pump always filled with liquid. Priming is not necessary. The liquid from the delivery pipe is guided by the chute 6, either directly to go to the suction sump through the auxiliary pipe 4 or to fall into the delivery tank 5, for flow measurement and then transferred to the suction sump for recirculation. Delivery pressure (p) is measured by the delivery gauge 3, fitted at the delivery pipe and vacuum (V) is measured by the vacuum manometer 2 connected to the suction pipe. The total head developed by the pump is the sum of the pressure and vacuum. In order to ensure correct readings of gauges, it is essential that there must be a minimum straight length of 4 to 6 times the diameter of the pipe before and after the gauges. The same condition should be followed for flow rate measuring instruments also. Watt-meters are connected to the electric prime mover for power measurement. Speed is measured by a tachometer.

The radial type centrifugal pumps are always started with gate valve under closed condition. For mixed and axial flow pumps and for all positive displacement pumps, pumps are always started with gate valve under fully opened condition. This is done in order to ensure minimum power consumption at the time of starting of pumps as well as for pump and prime move safety against overload. Tests are conducted for different flow rates by controlling the gate valve from full closed to full open positions.

At each position, total head, the flow rate and the power are measured. The speed is maintained constant throughout the test. The readings are entered in a tabular form (Table 3.1) and the performance of the pump is plotted in the form of a graph (Fig. 3.2).

Table 3.1: Pump test–development/production

Pump data **Type:** **S No:**	**Test Data** **Delivery type** **Wattmeter/Constant**	**Pressure gauge:**	**Test no:** **Date:**
Size: H = m Q = lps n = rpm Impeller Details: Casing Details:	Size of the Collecting Tank: Suction Lift:	Vacuum Gauge: Z = m	Time Tested by:

Sl. No
PGR(kGf/cm^2) Hg/cm^2)
Delivery Head (m)
VGR (V) mm of Hg
Suction Head (m)
Gauge Distance (m)
Vel Head Corr (m)
Total Head (m)
Rated Head (m)
Time for 20 cm sec
Qty (Q) lps
Rated (Q)
Voltage (V)
Current (A)
Wattmeter (kW)
Rated input (kW)
Pump (kW)

Pump output (kW)
(P_o) Efficiency
Remarks

H : m, Q : lps, η %, Suction lift: (h_s)

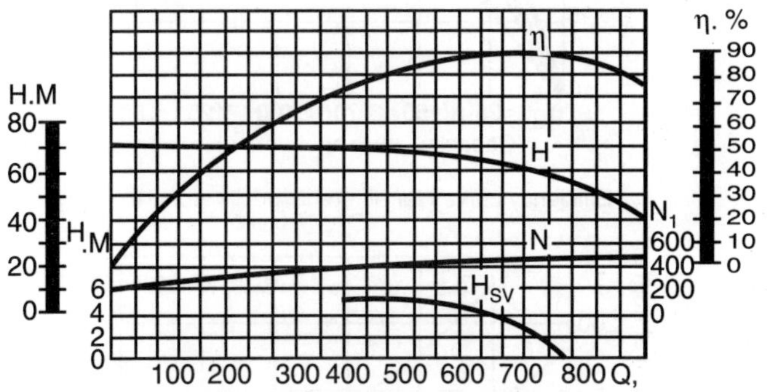

Fig. 3.2: Load test results

3.2 ANALYSIS OF PUMP PERFORMANCE

In Fig. 3.3, the performance of centrifugal pumps from radial to axial type are shown (H–Q N–Q, η–Q). It is seen from the graph, that when the specific speed increases the pump type changes from radial to axial. The ratio of shut off power to the power at optimum efficiency, changes from less than unity to 2½ to 3 times, when the specific speed of the pump increases. Referring to the discharge curve, the ratio of the shut off head to about 1.05, where as it is about 2½ to 3 times for axial flow pumps. For all optimum head, at maximum efficiency condition for radial flow pumps, is

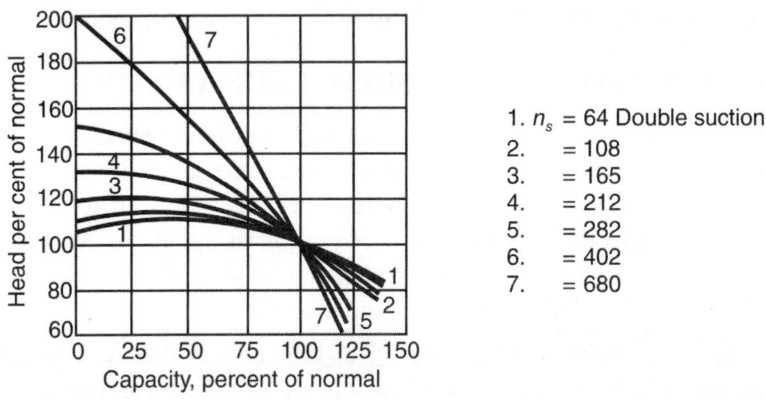

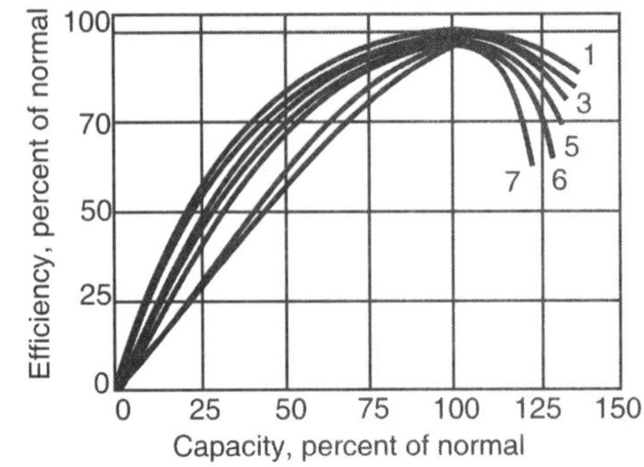

Efficiency curves for different specific speeds.

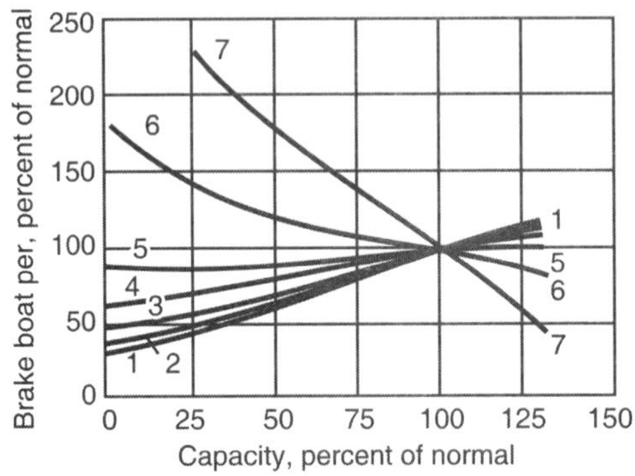

Fig. 3.3: Change of (*H-Q*), (*N-Q*), (η-*Q*) for different n_s

roto dynamic pumps, radial and axial flow pumps, range of safe operating region must be specified for non-overloading operation as well as for unstable flow conditions of operation. A drooping curve near partial flow conditions in radial flow pumps, and sudden change in head discharge curve in axial flow pumps indicate unstable region of operation.

3.3 UNIVERSAL CHARACTERISTICS OF A PUMP

In many occasions, pumps may be operated at different speeds. The test results for different speeds are plotted in the same graph (Figs. 3.4 and 3.5). Equal efficiency curves are plotted over this graph. In some cases, equal suction specific speed C or NPSH (σ) curves are also plotted on the graph. Such a graph is called universal characteristics, where in the entire performance, operating range of a pump is available.

Universal characteristics are also drawn for different impeller diameters, which are mostly used for chemical industries. This is done only for centrifugal pumps of radial type. In axial pumps, universal characteristics are drawn for different impeller vane settings.

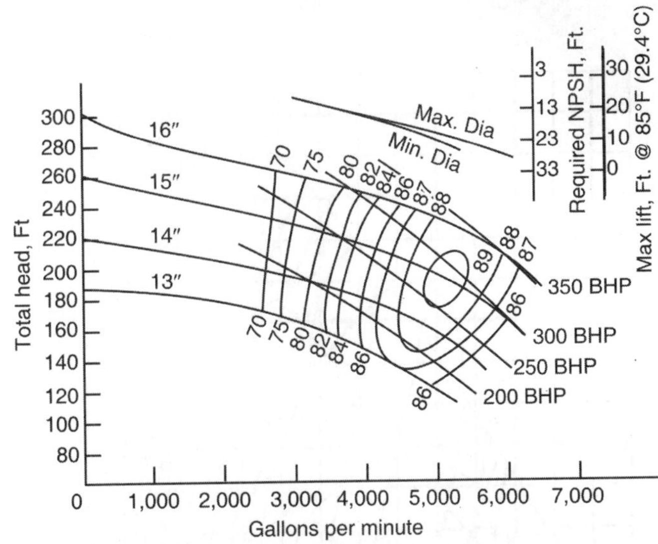

Fig. 3.4: Universal characteristics of a radial type centrifugal pump

3.4 THEORETICAL HEAD–DISCHARGE (*H*–*Q*) CURVE

The theoretical head-discharge curve is a straight line inclined at an angle of β_2 (outlet vane angle) from the Y-axis (Fig. 3.6). The ordinate will be (u_2^2/g). Owing to the use of finite number of vanes, the actual shut-off head will be lower and equal to $0.585\ u_2^2/2g$ (line 2). When flow rate increases frictional and diffusion losses increase as per the law $h_f \ \alpha\ Q^2$. Due to this effect, *H-Q* curve changes to a form of inverted parabola. Further reduction in *H-Q* curve takes place, due to shock at the entrance and at the exit of the impeller. It follows the law, $h_{sh} \propto \alpha\ (C_u^2/2g)$. The final shape of the *H-Q* curve is the line 4.

In mixed and axial flow pumps, reversal of flow takes place especially at partial flow rates (change in absolute velocity) which is indicated by a steep rise in *H-Q* curve.

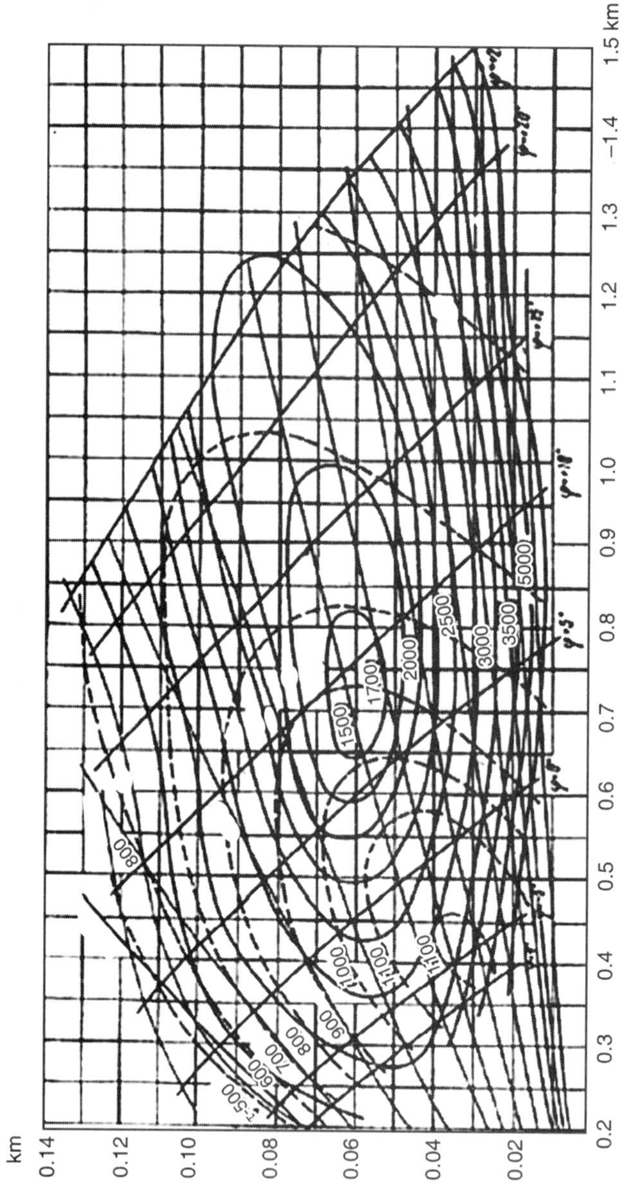

Fig. 3.5: Universal characteristics of an axial flow pump

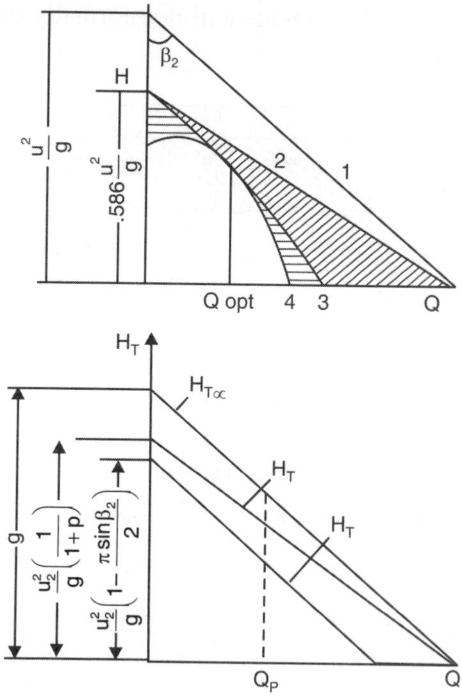

Fig. 3.6: Theoretical (H-Q) construction

3.5 ENERGY BALANCE IN PUMP

3.5.1 Energy Balance for Different Specific Speeds of Pump

In Fig. 3.12, energy balance for pumps of different specific speeds (n_s) are given. These are determined based on the experimental results on different pumps. Mechanical losses consist of stuffing box losses and bearing losses. Disc friction losses are created due to the impeller, as a disc, rotating in liquid in a closed chamber (volute casing). Volumetric losses are the leakage losses through clearance between wearing rings at both ends of the impeller, leakage through balancing holes and balancing drums, stuffing box and lantern rings, (only when pumping liquid is used for lantern ring). Hydraulic losses are the losses due to friction. When fluid passes through the impeller passage called eddy current losses or secondary losses or generally called as non-profile losses. Secondary losses occur and occupy major percentage of hydraulic losses at partial flow condition only. At optimum flow condition (higher efficient point in the *H-Q* curve) secondary losses will not normally exist. Only major friction losses prevail.

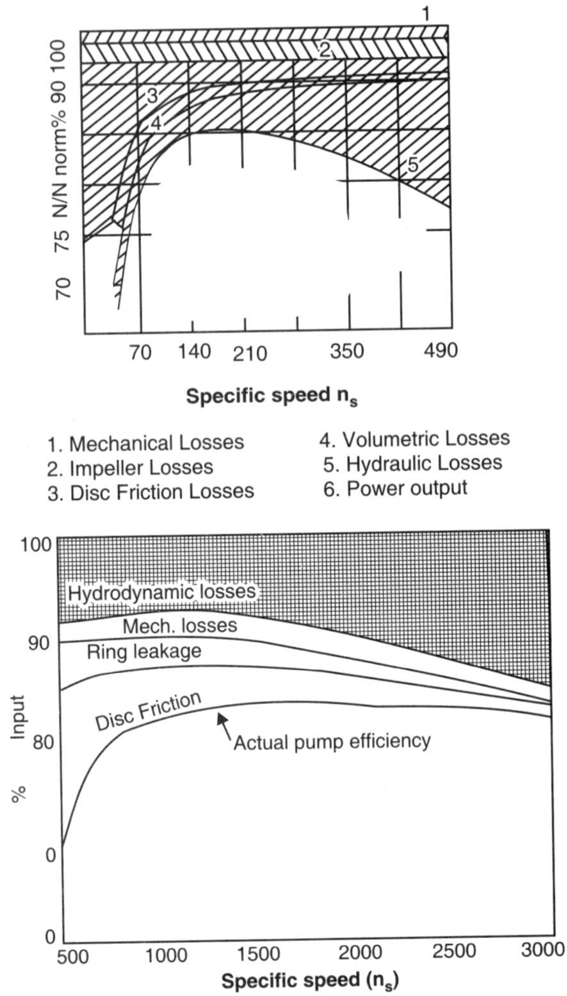

Fig. 3.7: Energy balance for different n_s

Mechanical and volumetric losses can be determined accurately by equations. These equations are obtained based on experimental results with theoretical analysis. In practice, knowing the over-all efficiency from pump tests, the hydraulic losses, profile, non-profile and shock losses, are determined graphically or analytically.

Hydraulic losses in casing are constant up to $n_s = 140$ and increases considerably at higher specific speeds. The efficiency reduction at lower specific speeds is mostly due to higher volumetric and disc friction losses.

3.5.2 Energy Balance of a Pump at Constant Speed

Energy balance for a pump $n_s = 135$ running at constant speed, but with different flow rate, is given in Fig. 3.8. Increase in volumetric losses is due to the higher pressure difference across the wearing rings, increase in mechanical losses is due to increased disc friction losses at reduced flow rate, which reduces the efficiency of the pump. However, major reduction in efficiency at reduced flow rate is caused by re-circulation. At higher flow rates, overall efficiency reduction is attributed, mostly, to hydraulic losses, non-profile and shock losses.

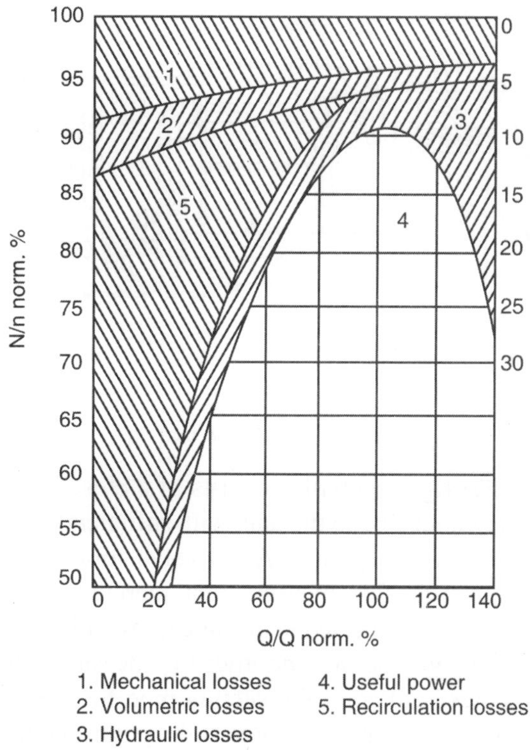

Fig. 3.8: Energy balanced for a pump

3.6 CAVITATION TEST

Cavitation test is carried out in a closed test rig. These tests cannot be carried out in load test rigs. However load test can be carried out in cavitation test rig (Fig. 3.4).

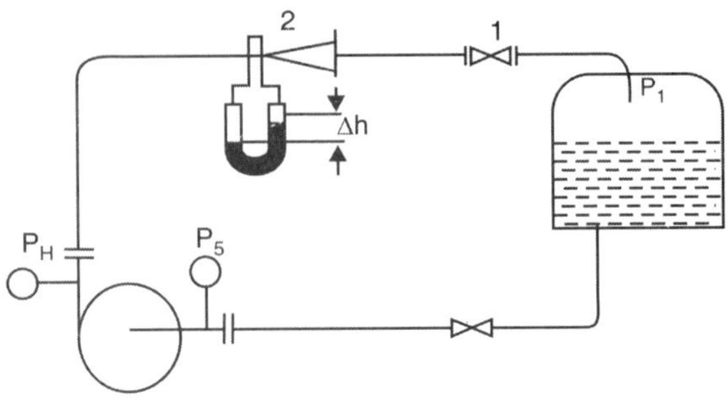

(a) Schematic diagram of a cavitation test rig

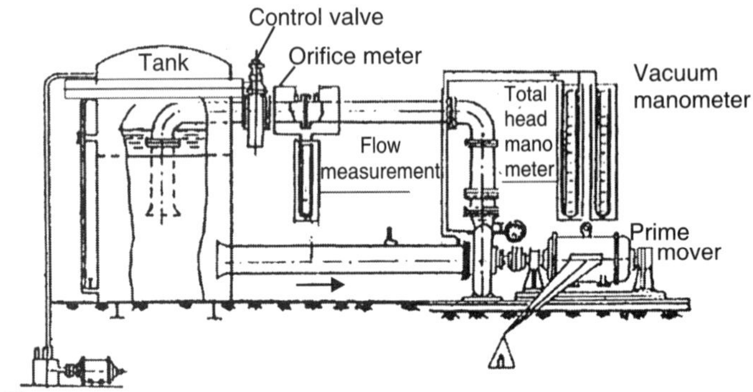

(b)

Fig. 3.9: Cavitation test rig

Cavitations test rig essentially consists of a closed tank. Suction and delivery pipes are connected to the tank with sufficient straight lengths for accurate measurements of all values. A gate valve, for flow control, a venturi meter or orifice meter for flow measurement, mercury manometer for total head measurement, are connected to the pipe lines with sufficient upstream and downstream lengths before and after the measuring points. Another mercury manometer is fitted at the suction pipe for the measurement of positive or negative (vacuum) pressures at the pump suction.

The closed tank is connected to a vacuum pump, by which the pressure in the tank can be altered to any desired value. The tank is filled with water up to 3/4 of its height and acts as suction and delivery reservoir.

The pump, to be tested is mounted on a base and connected to the prime mover, usually a dynamometer. Suction and delivery pipes are connected to the pump and all joints and stuffing box chamber are made air tight against any leakage with proper internal sealing arrangements.

A thermometer is fitted at the water line very near to inlet of the pump to measure the temperature of the water.

3.6.1 Method of Conducting a Cavitation Test

Cavitation test is conducted as follows: Select 6 to 7 points from the load test result of the pump, where NPSH value has to be determined. Usually these points will be 2 to 3 points before best efficiency point, 2 to 3 points after best efficiency point

1. Adjust the flow rate by operating the gate valve, to anyone of the points from the selected values. Note the total head, suction head, flow rate and the power. These will be same as in the load test results. Calculate efficiency (η), plot the readings in the graph, H, Q, N, $n = f\,(h_s H_{sv})$, which is the vacuum gauge reading.
2. Operate the vacuum pump and reduce the suction pressure in the tank, which simultaneously increases the vacuum at the pump entrance. Maintain the vacuum constant. Note down the readings of total head vacuum head, quantity, power, efficiency and speed. Plot these points in the same graph for the new vacuum head.
3. Increase the vacuum further by operating the vacuum pump and maintain it constant. Repeat the procedure until maximum possible vacuum is attained. When vacuum is increased further and further, at one point, there will be a sudden drop in one or all parameters (head discharge, power, efficiency) and all these values fluctuate. The pump produces noise and vibration, due to cavitation and erosion at the inlet edge of the impeller. This point is the critical point and further, vacuum cannot be created. Experiment cannot be proceeded further. One such test for a particular head-discharge condition is shown in Fig. 3.10. The limit of cavitation free operation is determined as 1% lower than the value of the suction head, where fluctuation starts.
4. The experiment is repeated for different points of head-discharge and at each point limiting suction head or NPSH values are determined (refer equation 1.7).
5. A combined NPSH vs Flow rate curve is now plotted on the same load test curve Fig. 3.10. This curve represents H_{sv} (min) for

each flow rate. The minimum most point of NPSH in the NPSH – discharge curve is the best operation point for cavitation-free operation. It is to be mentioned that best efficiency point and best NPSH points are not same and efficiency point and best NPSH points need not occur at the same discharge.

3.6.2 PARTS AFFECTED BY CAVITATION

In pumps, inlet edge of the impeller and corresponding surface in the casing are the most vulnerable places for cavitation damage. The volute tongue is also subjected to severe cavitation erosion. In some of the pumps, operated with very low total head and with high vacuum at suction, the outlet edge of the vane is also affected (Fig 3.7).

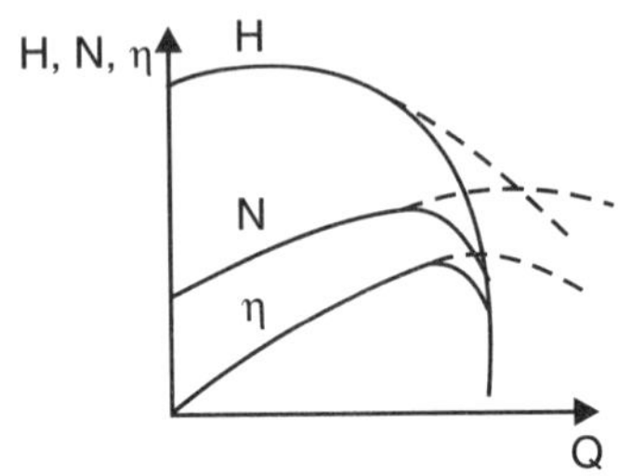

Fig. 3.10: Cavitation test result on performance

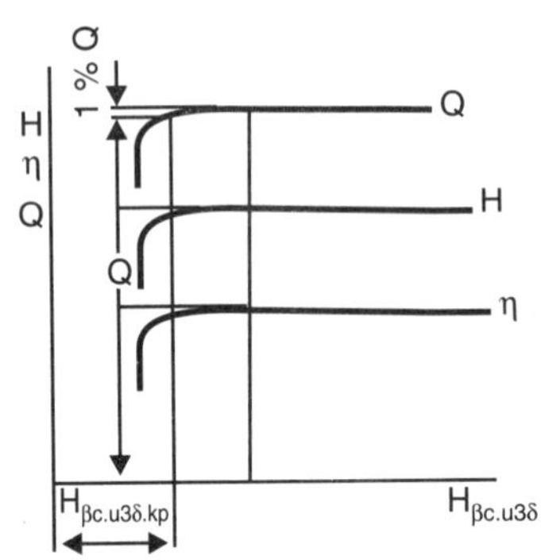

Fig. 3.11: Effect of cavitation

In mixed and axial flow pumps, the leading edge at the inlet of the vane is badly affected by cavitation. In positive displacement pumps, the suction ports and valves, inlet valve side of the cylinder or the cylinder liner and upper portion of the piston are mostly affected by cavitation.

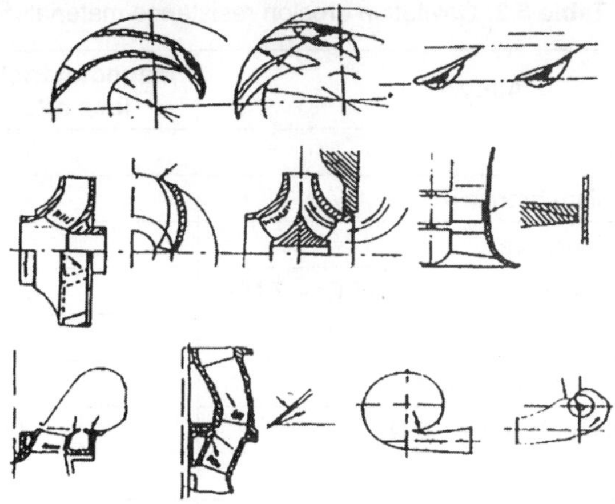

Fig. 3.12: Places affected by cavitation

3.6.3 Influence of Pump Parameters on Cavitation

1. Reduction in inlet edge thickness and rounding off the inlet edge improves considerably. Experimental analysis shows that reducing the inlet-edge, the thickness ratio (δ/D) from 0.923 to 0.154 improves the cavitational property by 200% at all flow rates.
2. Selection of optimum number of vanes improves cavitations performance. Increasing or decreasing number of vanes deteriorates NPSH value by about 150 to 200%. Efficiency is considerably reduced especially at reduced number of vanes and an unsteady flow was also observed.
3. Reduction in inlet area with constant breadth, increasing the hub diameter deteriorates cavitational property by about 50%.
4. Increase in inlet breadth improves cavitational characteristics.
5. Change in inlet vane angle in positive direction (increasing vane angle) does not influence cavitation. But change in negative direction reduces the cavitational property as high as 300%.

Table 3.2: Cavitation erosion resistance materials

Alloy	Magnetostriction weight loss after 2 h, mg
Rolled steelite	0.6
Welded aluminium bronze	3.2
Cast aluminium bronze	5.8
Welded stainless steel (2 layers, 17 Cr – 7 Ni)	6.0
Hot rolled stainless steel (26 Cr – 13 Ni)	8.0
Tempered rolled stainless steel (12 Cr)	9.0
Cast stainless steel (18 Cr – 8 Ni)	13.0
Cast stainless steel (12 Cr)	20.0
Cast manganese bronze	80.0
Welded mild steel	97.0
Plate steel	98.0
Cast steel	105.0
Aluminium	124.0
Brass	156.0
Cast iron	224.0

Despite the high resistance of this material to cavitation damage, it is not suitable for ordinary use because of its comparatively high cost and the difficulty encountered in machining and grinding.

Table 3.3: Cavitation erosion resistance of elastomeric coatings

Material	Sub type	Thickness of coating in rotating-disk cavitation tests, in (mm)	Cavitation test exposure period, h	Degree of erosion after exposure period, 150 ft/s (45.7 m/s)
Neoprene solvent base, brush-applied neoprene, cured sheet, cold-bonded neroprene, *in situ* cured and bonded polyurethane, liquid	A	0.030 (0.76)	24	Slight
	B	0.025 (0.68)	17	Slight
	-	0.062 (1.57)	14	None
	-			
	A	0.060 (1.52)	10½	None
	B	0.062 (1.57)	12	Slight
		0.018 (0.46)	12	Severe

Polyurethane cured	C	0.062 (1.57)	12	None
sheet, cold-bonded	A	0.060 (1.52)	14	None
Polysulfide, liquid	B	0.062 (1.57)	12	Severe
Polysiolane, liquid	-	0.062 (1.57)	12	Severe
Butyl, cured sheet, cold bonded	-	0.062 (1.57)	7	Severe
Butyl, *in situ* cured and bonded	-	0.060 (1.52)	2½	Severe
Cis-polybutadience (98%) cured sheet, cold bonded	-	0.060 (1.52)	12	Severe
Polybutadiene	-	0.060 (1.52)	10	None
(Polysulfide modified) in situ cured and bonded	-	0.060 (1.52)	13	Severe
Syrene-butadiene Copolymer, in situ cured and bonded (SBR)	-	0.060 (1.52)	24	None
Natural rubber, cured sheet, cold-bonded	-	0.062 (1.57)	10	None
Natural rubber, in situ cured, cold-bonded	-	0.060 (1.52)	16	Severe

4 Systems and Arrangement

4.1 HEAD MEASUREMENT IN A SYSTEM

The head required by the pump for pumping water from one tank to another depends upon the type of system. In Fig. 4.1, a simple system is shown. Let us assume that pressures at the suction and delivery reservoirs be p_s and p_d respectively and its levels from the reference line (datum) be Z_s and Z_d respectively. The line losses at suction and delivery pipes of the pump are designated as h_{fs} and h_{fd} respectively, whereas the gauge readings are p_1 and p_2 respectively. Let the gauges be fitted at a difference in level of X. A system consists of already laid pipelines, fittings, valves, collecting tanks (may be in closed or in open to atmospheric condition) in the field or at the work spot. The suitable pump is always selected to suit the system conditions. The discharge or the flow through the system, delivery, or suction connected to the pump will be same. Total head of a system (H_{sy}) is equal to the sum of the head of delivery system ($H_{sy}(d)$) and head of the suction system $H_{sy}(s)$ *i.e.*, $H_{sy} = H_{sy}(d) + H_{sy}(s)$.

For a real fluid flow with friction,

$$H_{sy}(d) = \frac{p_d}{\gamma} + Z_d + \frac{C_d^2}{2g} + h_{fd} \text{ (Energy required at delivery) and}$$

$$H_{sy}(s) = \frac{p_s}{\gamma} + Z_s + \frac{C_s^2}{2g} + h_{fs} \text{ (Energy available at suction)}$$

$$H_{sy} = H_{sy}(d) - H_{sy}(s)$$

$$= \left(\frac{p_d}{\gamma} + Z_d + \frac{C_d^2}{2g} + h_{fd}\right) - \left(\frac{p_s}{\gamma} + Z_s + \frac{C_s^2}{2g} + h_{fs}\right)$$

Since $(Z_d + Z_s) = +(h_d + h_s)$ and taking $h_f = h_{fs} + h_{fd}$

$$H_{sy} = \left(\frac{p_d - p_s}{\gamma} + h_s + h_d + h_f + \frac{C_d^2 - C_s^2}{2g} \right) \qquad (4.1)$$

Suffix '*d*' indicates delivery, '*s*' indicates suction , '*sy*' indicates the system, 'h' static head or height, h_f – frictional loss of head. If the delivery and suction pipe diameters are equal then $C_d = C_s$ where

$C = \frac{4Q}{\pi d^2}$. Hence $\left(\frac{C_d^2 - C_s^2}{2g} \right)$ = 0. If the delivery and suction tank(s) are open to atmosphere then $P_d = P_s = P_{atm} = 0$ (open system)

The frictional loss of head

$$h_f = \frac{flv^2}{2gd} = \frac{fl\ 16Q^2}{\pi^2 d^4 2gd} = \frac{8flQ^2}{\pi g d^5} = 0.0826 \frac{flQ^2}{d^5} \qquad (4.2)$$

Since for a system '*f*' the frictional coefficient, '*l*' length of the pipe, '*d*' the diameter of the pipe are constant so,

$$\left(h_f + \frac{C_d^2 - C_s^2}{2g} \right) = \text{Constant } Q^2 \qquad (4.3)$$

Also for a system $H_{st} = \frac{p_d - p_s}{\gamma} + h_s + h_d$ and it is constant for the given installation or site or field conditions. Hence, $H_{sy} = H_{st} + \text{const } Q^2$. If a graph between H_{sy} and Q is drawn, it will be a parabola (as shown in Fig. 4.1) with H_{st} at Y-axis ($Q = 0$).

A system, in a pumping plant, consists of suction and delivery pipe lines along with all fittings such as sudden or gradual reducer or expander, all types of valves, such as gate valve, non-return valve, butterfly valve, tee or cross joints etc. A system is an already available pipe layout in the field to suit the actual conditions; for example, city water supply, multi-storey buildings, layout in chemical fertilizer and in power station etc. It consists of different lengths and diameters or pipes, fitted with various pipe fittings.

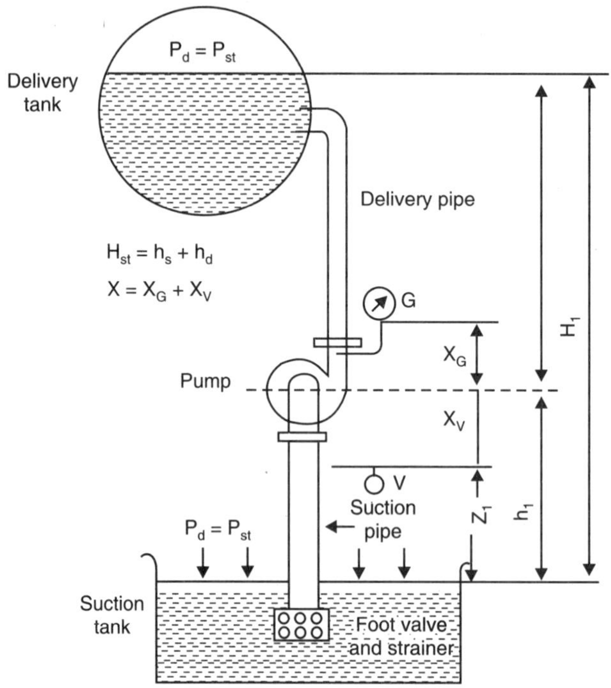

Fig. 4.1: Pump with a system

The level between inlet and outlet end of the system may be equal or different. Total resistance offered by a system is the sum of major and minor losses in the system along with the level difference between inlet and outlet end of pipe.

$H_{sy} = h_s + h_{fs} + h_d + h_{fd} + \Sigma h_{fmi}$ for all fittings attached to this pipe line.

Suffix *s* is for suction and *d* is for delivery. Major friction losses in pipe are expressed as $h_f = \dfrac{flv^2}{2gd}$ and minor losses are expressed as $h_{fmi} = K\dfrac{v^2}{2g}$, where *K* is the constant for that particular fittings attached to the main pipe.

$$\therefore \quad H_{sy} = (h_s + h_d) + \frac{fl_s v_s^2}{2gd_s} + \frac{fl_d v_d^2}{2gd_d} + \Sigma K_s \frac{V_s^2}{2g} + \Sigma K_d \frac{V_d^2}{2g} \qquad (4.4)$$

Since $V = \dfrac{Q}{A} = \dfrac{4Q}{\pi d^2}$ and flow rate *Q* is constant through both pipes

$$H_{sy} = (h_s + h_d) + \frac{16fl_sQ^2}{2g\pi^2 d_s^5} + \frac{16f_dQ^2}{2g\pi^2 d_s^5}$$

$$+\Sigma(K_s)\frac{16Q^2}{2g\pi^2 d_s^4} + \Sigma(K_d)\frac{16Q^2}{2g\pi d_s^4} \quad (4.5)$$

In equation (4.4) the values l_s, l_d, l_s, l_d are constants for the already laid piping system, g, π, f are constant. Hence, it can be expressed as

$$H_{sy} = H_{st} + (K_{sm} + K_{dm} + K_{mis} + \Sigma K_{mid})Q^2 = H_{st} + KQ^2 \quad (4.6)$$

where, $K = K_{sm} + K_{dm} + \sum K_{mis} + \sum K_{mid}$

$$K_{sm} = \frac{16fl_s}{2g\pi^2 d_s^5}, K_{dm} = \frac{16fl_d}{2g\pi^2 d_d^5}, K_{mis} = \frac{K_s 16}{2g\pi^2 d_s^4}, K_{mid} = \frac{K_d 16}{2g\pi^2 d_d^4}$$

and static head (level difference between inlet and outlet points in a pipe line) $H_{st} = (h_s + h_d)$.

Equation (4.6) represents a parabola (Fig. 4.2) in H_{st}–Q coordinates. If $H_{st} = 0$, the parabola starts from origin. Parabola will be more curved upwards if length of pipeline is increased, pipe diameter is reduced, pipe fitting are increased, or pipe fittings are changed, gate valve partly opened, change of gate valve, with non-return valve or globe valve etc. which totally increase the value 'K' for minor losses. If the system arrangement is reversed the parabolic curve comes down.

This parabolic curve for any system is an already available graph since all values are already available from site conditions, and constant K from reference manuals.

4.2 Pump System Curve Construction

In Fig. 4.3, a pumping system working between suction tank *A* and the delivery tank *B*, is at a higher level than tank. Both the tanks are above the pump level *i.e.*, the system in suction side is with positive suction (suction head). OE is the system curve for the suction pipe and is drawn in the reverse direction, below the water level in the suction tank (*A*). CS is the system curve for the delivery pipe and is drawn above the water level in the delivery tank *B*. The vertical distance between *CD* and *OE* is the total system head for the given flow rate point *Q*. Curve *CF* is the total system curve (*i.e.*, *OE* + *CD*). The meeting point of this curve with the (*H-Q*) curve of the pump at *A* is the operating point. The vertical distance H_s gives the manometer reading in suction pipe and the vertical distance H_d is the monometer reading in the delivery pipe.

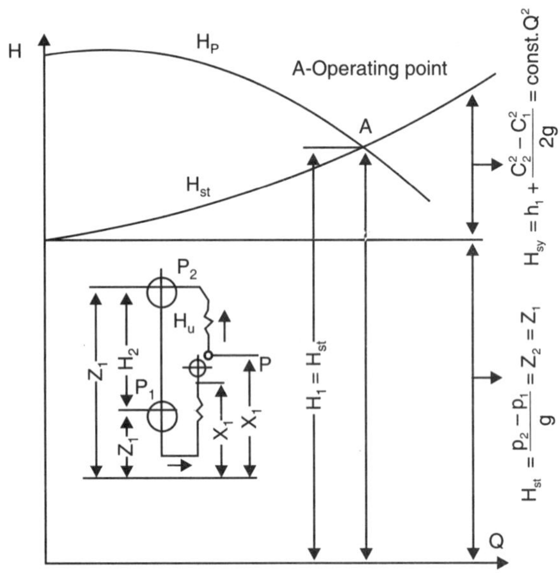

$X_2 - X_1 = X,\ h_1 = h_f = h_{fs} + h_{fd}$

Fig. 4.2: Pump operating condition pump head H_p and system head $H_{sv} = H_{st} + H_{dy}$

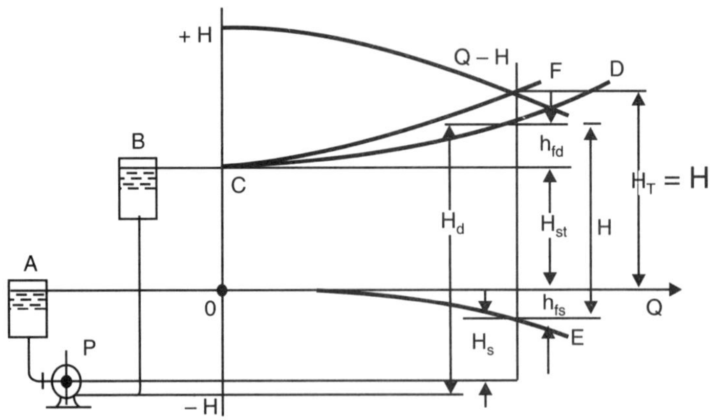

Fig. 4.3: System head for positive suction head

4.2.1 Systems in Series

If two systems *A* and *B* are kept in series, then total resistance of the system is the algebraic sum of individual resistances of system *A* and *B*, since same quantity of flow passes through both pipes.

$$Q_T = Q_A = Q_B \text{ and } H_{SYT} = H_{SYA} + H_{SYB} \qquad (4.7)$$

In Figs. 4.4 and 4.5, Equation (4.7) is illustrated in a graphical form.

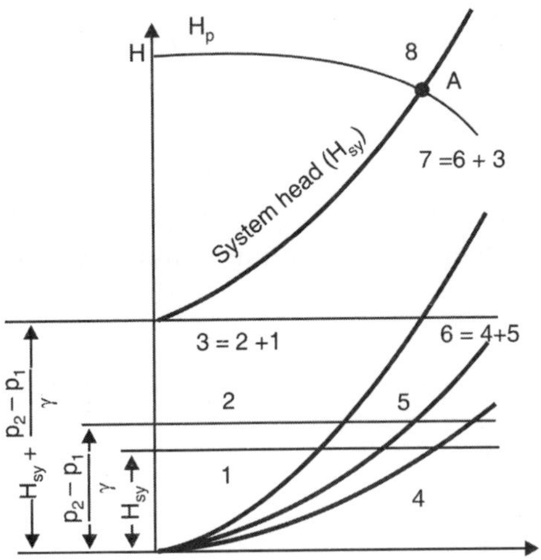

1. Static height between delivery and suction
2. Delivery and suction tanks pressure difference
3. Total static head (H_{st}) = (2 + 1)
4. Delivery and suction velocity head difference
5. Delivery and suction pipe friction losses ($h_{fd} + h_{fs}$)
6. Total dynamic head (H_{dy})
7. Total system head $H_{sy} = h_{st} + h_{dy}$ i.e., ($H_{sy} - Q$)
8. Pump head (H_P) i.e., ($Hp - Q$)

Fig. 4.4: H_{SY}- system head construction

A number of Q lines at frequent intervals, from origin to Q_{max} parallel to Y-axis *i.e.*, Q = constant lines are drawn in the graph having $H_{SY (A)} - Q$ and $H_{SY (B)} - Q$ graphs (Fig. 4.5), the two ordinate lengths are added to get the point of $H_{SYT} = H_{SYA} + H_{SYB}$ for one value of Q. Similarly other, H_{SYT} points for other Q are determined. A smooth curve joining these H_{SYT} points gives $H_{SYT} = f(Q)$ curve for combined series operation of two systems. The same procedure is followed for many systems connected in series based on the equation (4.7). If the *H-Q* curve of the pump is overlapped on the combined system graph, point A is the operating point, when both systems work. Several systems operating in series is equal to one system with higher system head.

4.2.2 Systems in Parallel

If two systems 1 and 2 are operated in parallel (Figs. 4.7 and 4.9) the total quantity (Q_T) is the sum of individual quantities passing through each systems and the total resistance is equal to individual resistance of each system *i.e.*,

$$Q_T = Q_1 + Q_2 \text{ and } H_{SYT} = H_{SY_1} = H_{SY_2} \tag{4.8}$$

Many horizontal lines, *i.e.*, H = constant lines, are drawn at frequent intervals from origin to shut off head, in the graph where $H_{SY_1} - Q$ and $H_{SY_2} - Q$ curves are drawn. Individual horizontal distances between Y-axis and H_{SY_1} and H_{SY_2} are added together to get $Q_T = Q_1 + Q_2$. The same procedure is followed for many systems in parallel operation *i.e.*, $Q_T = Q_1 + Q_2 \dots Q_n$ and $H_{SYT} = H_{SY_1} = H_{SY_2} = H_{SY_3} \dots = H_{SY_n}$ In Fig. 4.6, systems operation is parallel is illustrated.

If H–Q curves of the pump is inserted on this combined graph. Operating points where both systems are in operation is point A and operating points when one any system is in operation is indicated by points 2 and 3.

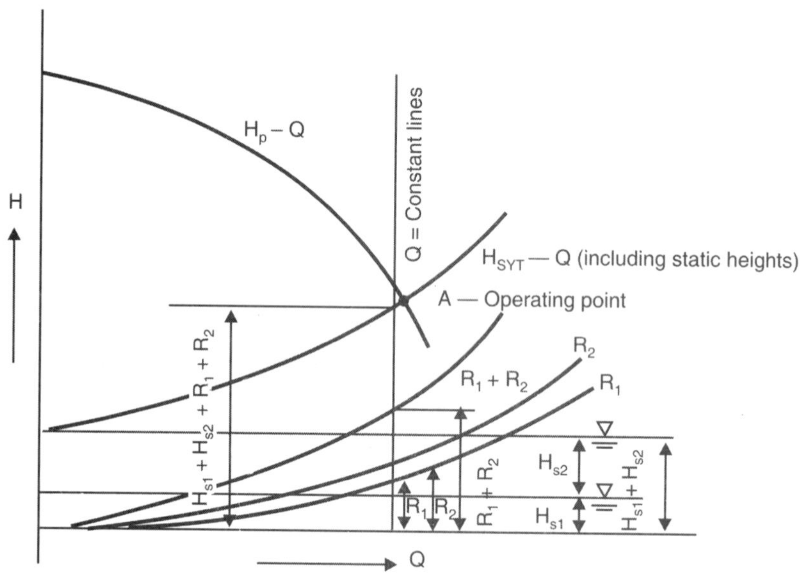

$H_{SYT} = H_{SYA} + H_{SYB}$, $H_{SYA} = H_{s_1} + R_1$, $H_{SYB} = H_{s_2} + R_2$

Fig. 4.5: Two systems in series

4.2.3 Operation of Pump in Parallel and with the System Characteristics

If two pumps 1 and 2 are operated in parallel, the quantity of flow will be the sum of individual quantity of pumps. Both these pumps will be operating at the same head (Figs. 4.8 and 4.9) (1 + 2). If the piping characteristics (3) is plotted on the same graph, the operating point (3) again will be the meeting point of the curves (1 + 2) and (3). Parallel system of operation of pumps are used for larger quantity at the same total head.

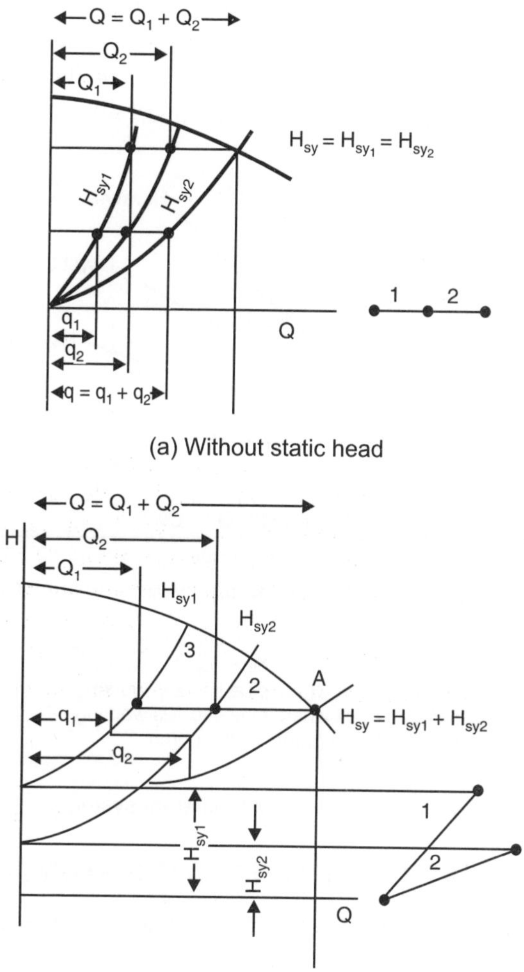

(a) Without static head

(b) With different static head

Fig. 4.6: Pump with two systems and operating points

4.2.4 One Pump with many Systems in Operation

Curve R_1 is the system characteristics for the first pipe line and curve R_2 is the system characteristics for the second pipeline. The combined system curve will be curve $(R_1 + R_2)$ and is obtained by adding individual flow rates for the same head (Fig. 4.8).

If the pump performance (*c*) is plotted on the same graph, the meeting point (X) will be the operating point of the system. The quantity of flow in each pipe will be determined by drawing a line from point (X) parallel to X–axis. The meeting points of this line with the curves A and give the individual flow rates. If a number of pumps either in series or in parallel are connected to a number of system graph C will be the combined pump graph and graph A + B will be combined graph of all systems.

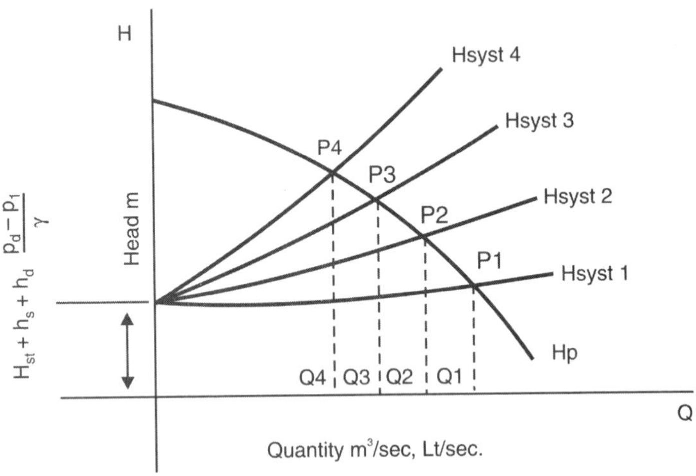

P4 – Operating point when pipe line 4 is working.
P3 – Operating point when pipe line 3 is working.
P2 – Operating point when pipe line 2 is working.
P1 – Operating point when pipe line 1 is working.
Hp-Head – discharge graph for the pump.

Fig. 4.7: One pump with different systems

4.3 DETERMINATION OF OPERATING POINT

H-Q curve of the pump to be connected to the system (between suction and delivery lines, *i.e.*, point '*s*' and '*d*') is overlapped on the available H_{SY}-*Q* curve, the meeting point 'A' is the point of operation of the pump. Point A changes for different *H-Q* curve of the pump, for example, *H-Q* curve for different speeds, with the same H_{SY}-Q curve or with the same pump but with different (H_{SY}-Q) curve (*i.e.*, system curve with different fittings). Head and discharge at point A will be the operating parameters. This parameter need not be the optimum parameter of the pump. Power consumption and efficiency can be obtained from N_i-Q and η-Q curves for the pump at point 'A'.

4.4 COMBINED OPERATION OF PUMPS AND SYSTEMS

4.4.1 Pumps in Series

Figure 4.8, two pumps are operated in series, then the total quantity is equal to the individual quantity of each pump. The same quantity passes through all pumps. Total head is the sum of heads developed by individual pumps *i.e.*,

$$Q_T = Q_1 = Q_2 \text{ and } H_T = H_1 + H_2 \tag{4.9}$$

Same procedure is followed for many pumps operated in series *i.e.*,

$$Q_T = Q_1 = Q_2 = \ldots\ldots = Q_n \text{ and } H_T = H_1 + H_2 + \ldots\ldots\ldots + H_n$$

In order to get the combined $H_T = f(Q)$ for combined operation in series, a number of Q = constant lines are drawn in the graph where individual H_1-Q and H_2-Q curves are already drawn. Total head of all pumps running together H_T is obtained by adding individual total heads as H_1 and H_2 for the same Q. A number of points so obtained is joined together by smooth line, which gives H_T-Q, for two pumps in series.

If system line (H_{sy}-Q) is overlapped on this graph, point 'A' is the operating point when both pumps are operated in series. Point A is the operating point when individually any one pump is operated. Point 'A' will be the operating point, when only one system (instead of both) is operated with one pump.

4.4.2 Pumps in Parallel

If two pumps are operated in parallel, then both pumps work under the same head. Total quantity (Q_T) is the sum of the quantities of the individual pumps *i.e.*,

$$H_T = H_A = H_B \text{ and } Q_A + Q_B$$

A number of H = constant lines are drawn in the graph where individual (H-Q) curves of pump C and D are available. For one value of

H = Constant line, Q_A and Q_B lengths are added together to get Q_T. In the similar manner a number of H = constant points are selected and Q_T is determined. A smooth line joining all these Q_T points give the combined graph ($H_T - Q$).

If a system curve ($H_{SY} - Q$) is overlapped on the combined ($H_T - Q$) of the pump and individual ($H_A - Q$) and ($H_B - Q$) graphs of the pumps, the operating point 3 is when both pumps operate and points 1 and 2 are when any one pump is operated.

Same procedure is followed for many pumps when operated in parallel.

$$H_T = H_A = H_B = \ldots\ldots\ldots\ldots = H_n \text{ and } Q_T = Q_A + Q_B + \ldots. + Q_n$$

Regulation of pumps in series or parallel is also achieved by sudden switching off of one or two pumps. This method saves the power requirement to the pump.

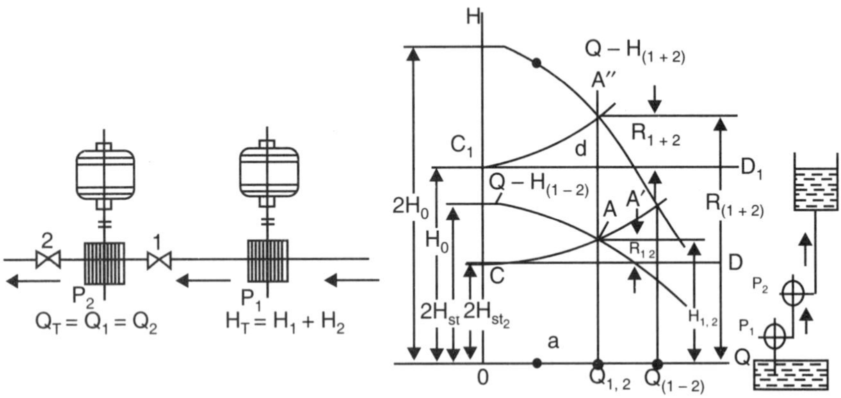

Fig. 4.8: Pumps in series (*H-Q*) curve

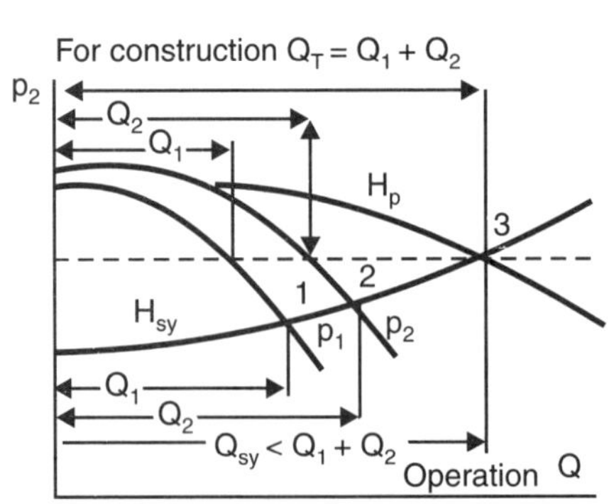

Fig. 4.9: Two pumps in parallel $Q_1 + Q_2 = Q_T$, $H_1 = H_2 = H_T$

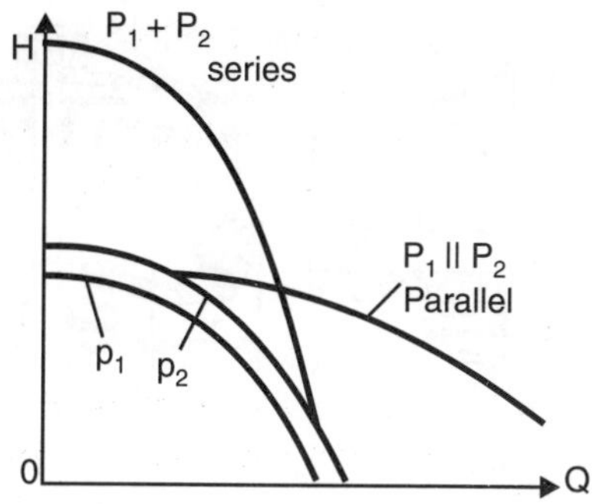

Fig. 4.10: Comparison of series and parallel operation

4.5 STABLE AND UNSTABLE OPERATION IN A SYSTEM

4.5.1 Stable Operation

In Fig. 4.12 $(H - Q)$ curve of pump meets $(H_{SY} - Q)$ curve of a pumping system at two operating points A and point B.

At both point A and B, condition $H_{SY} = H_p$ is prevailing. But, actually point A is the stable operating point and point B is the unstable operating point. Let the quantity at point A be shifted to the higher side of the quantity by a smaller amount (ΔQ_A). At the new point $H_{SY} > H_p \Delta H = H_{SY} - H_p > 0$. Pump supplies quantity at a lower pressure (H_p), where as the system is at a higher pressure (H_{SY}). Pump cannot supply flow against high pressure of the system. The flow gradually reduces until pump pressure (H_p) is equal to system pressure (H_{SY}). Same situation occurs when flow is reduced by a small amount *i.e.*, $(-\Delta Q_A)$. Now, $H_p > H_{SY}, \Delta H = H_{SY} - H_p < 0$. Pump supplies water at a higher pressure than the system pressure. This difference in pressure causes an increase in quantity. This increase in quantity reduces the pump pressure (H_p) and increase the system pressure (H_{SY}), until both the pressures are equal *i.e.*, until point A is attained. Entire process is automatically carried out. Pump can operate only at point A where the condition is $H_p = H_{SY}$. Hence, point A is a stable operating point.

4.5.2 Unstable Operation

Normal head-discharge curve for a centrifugal pump will have a shape of gradually rising characteristics, when quantity of flow is reduced from maximum to zero flow condition (Fig. 4.12). In some of the pumps due to heavy secondary losses at very flow rates, head-discharge curve of the pump droops down instead of raising after attaining maximum head at

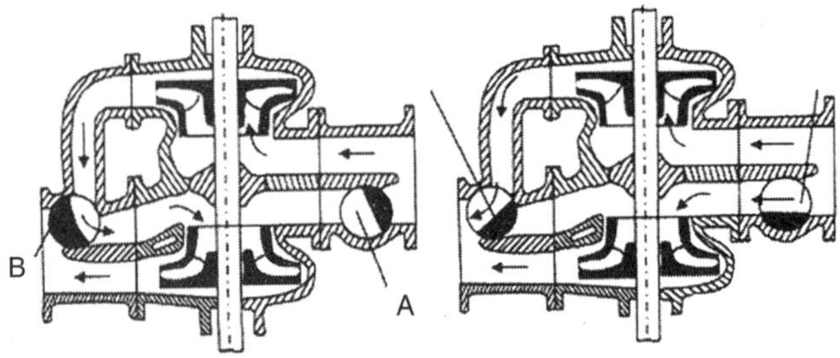

(a) in series (b) in parallel

Fig. 4.11: Schematic diagram for two pumps to operate (a) in series and (b) in parallel

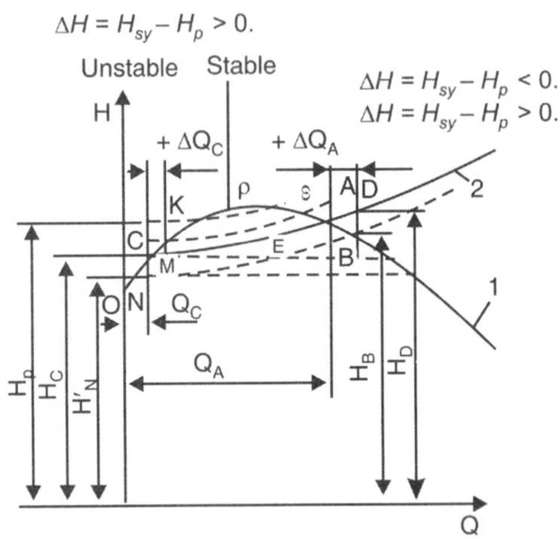

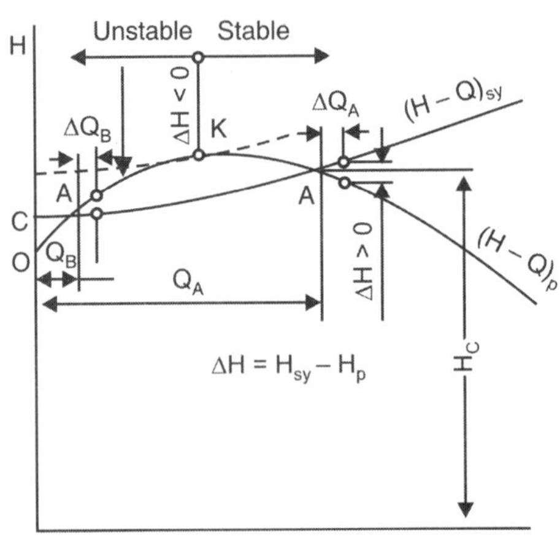

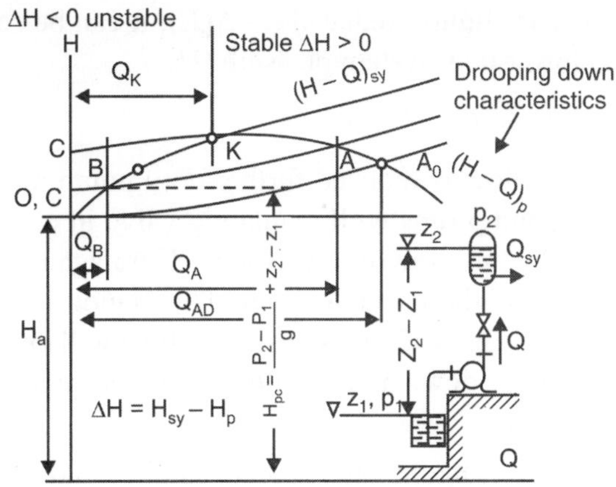

Fig. 4.12: Stable and unstable operation in pumping system

at certain quantity (point K) in Fig. 4.12. Always shut off head ($Q = 0$) will be the highest head in ($H_p - Q$) curve for all pumps of good design. But due to high secondary flow at low flow rate, ($H_p - Q$) curve droops down and the shut off head will be lower than the normal value.

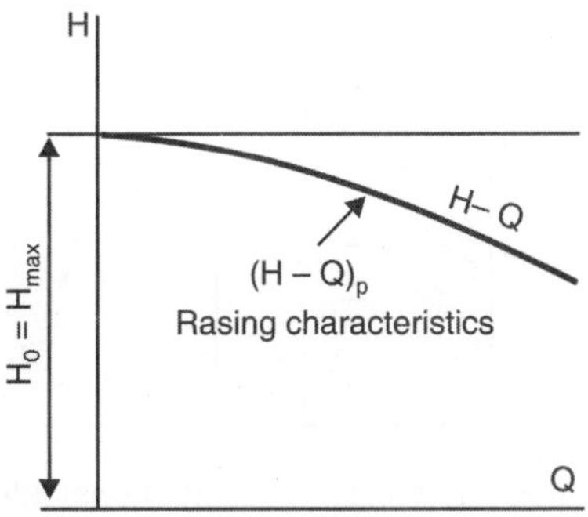

Fig. 4.13: Condition for stable operation $H_0 = H_{max}$

In Fig. 4.12, point B is the meeting point of the ($H_{SY} - Q$) and ($H_p - Q$) curves. A small increase in quantity ($+ \Delta Q_B$) moves the operating point to the right hand side of the graph ($+ Q$ direction). At this stage, system pressure H_{SY} is lower than the pump Pressure (H_p), $\Delta H = H_{SY} - H_p < 0$ (negative value) as a result pump supplies more quantity to the system, which makes the point to move further to the right hand side ($+ Q$ direction). Point B cannot be reached at all. Pump supplies more and more quantity to the systems until point A is attained.

If the quantity is slightly reduced ($-\Delta Q_B$), it can be seen that pump pressure (H_p) is lower than system pressure (H_{SY}).

$$\Delta H = H_{SY} - H_p > 0$$

The quantity of flow is reduced further until pump reaches the point (Q). At this stage, pump runs with regulating valve in opened condition, but there will not be any flow in the system. Here again point B cannot be reached. Point B is the unstable point. This region is called unstable region curve OBK in Fig. 4.12. This unstable effect will exist in all pumps if ($H_p - Q$) curve droops down at lower flow rates. The pump should not be operated in this region.

However, pumps can be operated at stable condition even when drooping down characteristics prevails at low flow rates. In Fig. 4.11, the shut off head of pump (at $Q = 0$) is lower than the shut off head of the system. In Fig. 4.12, pump shut off head is at a higher level than system shut off head. System curve and pump curve meet at point B, which is located at Unstable region of pump characteristics.

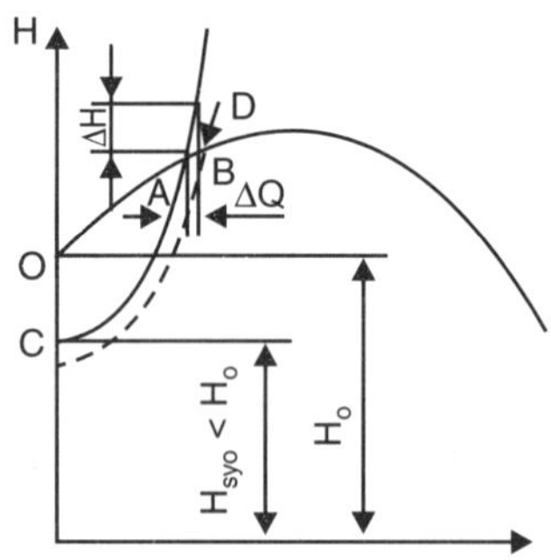

Fig. 4.14: Condition for stable operation at unstable region $H_o > H_{syo}$

The condition of point B in Fig. 4.15 is same as the condition of point A in Fig 4.12. Any small increase in $+\Delta Q_B$ increase H_{SY} and decreases H_P *i.e.*, $\Delta H = H_{SY} - H_P -> 0$ (+ve). Due to higher resistance in system, quantity reduces. This reduction continues until point B is reached. Any small decrease in flow rate ($-\Delta Q_B$) $H_P > H_{SY}$ or $\Delta H = H_{SY} - H_P < 0$. Higher pressure in pump increase the flow to the system until point B is reached. Therefore, point B is a stable point although is reached. Therefore, point B is a stable point although it lies on the drooping down side of pump characteristics. Mathematically, stable condition will exist if $\frac{dH_{SY}}{dQ} \quad \frac{dH}{dQ}$.

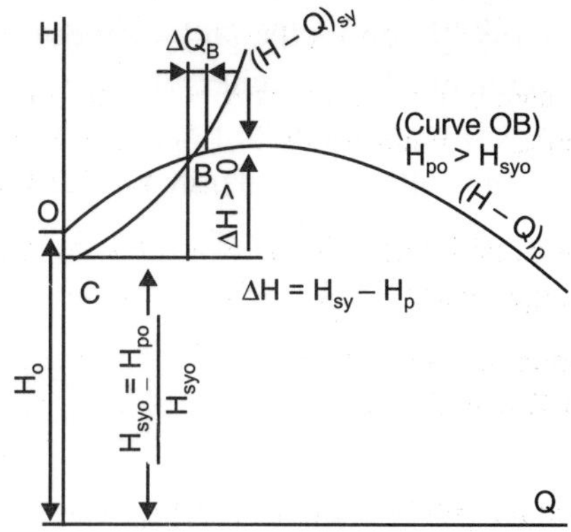

Fig. 4.15: Condition for stable operation in unstable region

An example of this condition is illustrated in Fig. 4.16.

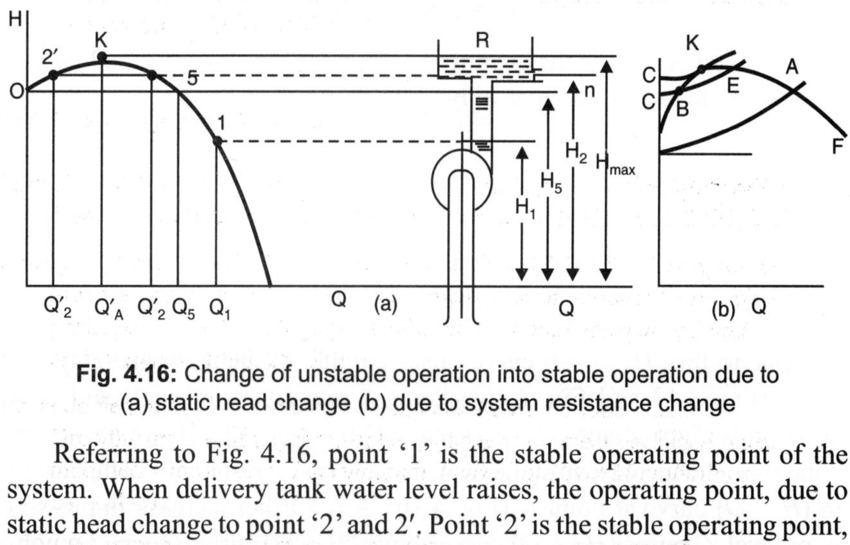

Fig. 4.16: Change of unstable operation into stable operation due to (a) static head change (b) due to system resistance change

Referring to Fig. 4.16, point '1' is the stable operating point of the system. When delivery tank water level raises, the operating point, due to static head change to point '2' and 2′. Point '2' is the stable operating point, whereas point 2′ is the unstable operating point. This unstable region starts from point '5' and above. Maximum height of water level in the tank can be up to point 'K'. Further increase in water level leads to reverse flow (Refer section 4.14).

4.6 REVERSE FLOW IN PUMP

In Fig. 4.17, a pumping system is illustrated where in the system *i.e.*, suction and delivery pipe, remain same. A closed delivery tank (condenser, accumulator) is attached to the system. Quantity Q_p enters the tank from pump whereas a quantity Q_1 leaves the tank to another system. Quantities Q_p and Q_1 are independent of each other. If $Q_p > Q_1$, water level in the tank gradually raises. Correspondingly, pressure p_2 also increases. Since system remains same, the pattern of dynamic part of system curve remains

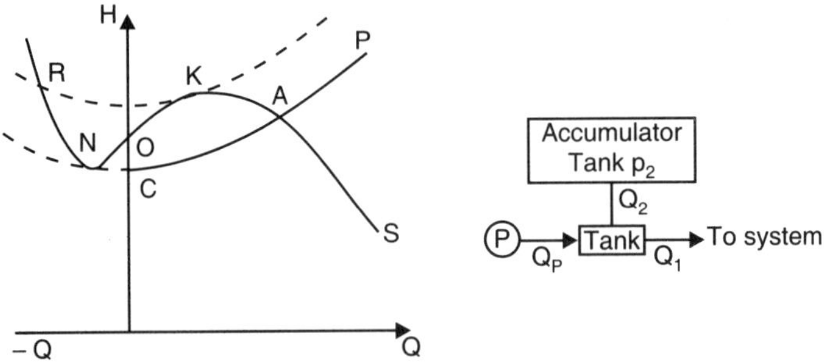

Fig. 4.17: Reverse flow in pump

same; whatever may be the quantity Q_2, increase in pressure p_2 and raise in level Z_2 in the tank. Change is only in the static head *i.e.*,

$$H_{sy} = \frac{p_2 p_1}{\gamma} + (Z_2 - Z_1) + \frac{C_2^2 - C_1^2}{2g} = H_{st} + H_{dy}$$

Where, Formula $H_{st} = \frac{p_2 p_1}{\gamma} + (Z_2 - Z_1)$ and $H_{dy} = \frac{C_2^2 - C_1^2}{2g}$

Due to this, $(H_{SY} - Q)$ curve raises parallel to the previous $(H_{SY} - Q)$ curve (Fig. 4.13 and Fig. 4.16). Point A moves towards left and point B towards right along $(H_p - Q)$ curve. Still $Q_A > Q_B$. However, at one condition, the increase in static head makes $(H_{SY} - Q)$ curve will be tangent to $(H_p - Q)$ curve at point K (Fig .4.16). Any further increase in pressure p_2 or level Z_2 raises $(H_{SY} - Q)$ curve above $(H_p - Q)$ curve. Operating point K moves to point R, where only further increase in H_{SY} is possible, *i.e.*, to the area of reversible flow. Flow moves from the tank to the pipe. This induces a reversible water wave. Pressure P_2 reduces. Operating point R moves to point N. Because flow reduction is possible only along the curve RN. Any further reduction in pressure moves the operating point from N to A and then gradually to S, because only at point S, further reduction in pressure is possible. Thus a cycle is completed. This cycle repeats so long as there is no change in the system. Pressure fluctuates at a faster rate. If the flow fluctuation is large, the unstable condition starts even earlier when operating point is at A. Even a small disturbance in pressure can induce unstable condition, if operating point is very near to point K.

Reverse flow prevails until $H_{st} = H_0$ (at $Q = 0$) *i.e.*, shut off head when $H_{st} < H_0$ pump starts pumping with operating point at A or below point A under stable condition. When $Q_A < Q_K$ stable operation of the pump is not possible. $\frac{dH_{SY}}{dQ} < \frac{dH_p}{dQ}$ from $Q = 0$ to $Q = Q_{K.}$ Hence unsteady flow prevails between these two points.

Quantity 'Q' passing through the system will be either $Q = 0$ or $Q = Q_K$. At this region reverse flow takes place. Drooping down characteristics at low flow rate is the unstable region so long $H_{st} > H_0$. It becomes stable when $H_{st} < H_0$ and stable operations starts from the condition $H_{st} < H_0$ to the condition $H_{st} = H_0$. The same thing will happen even pumps are running in parallel or in series. The only difference is that combined characteristics must be studied. However, a raising characteristics of $H - Q$ curve from Q_{max} to Q_0 *i.e.*, up to shut off head is the best $H - Q$ curve for stable operation.

Such pressure fluctuations quite frequency occurs in boiler feed water storage drums, condensers, accumulators and in pipes with elastic properties.

This effect is more predominate in pumping compressible fluids such as gas, air especially pumping to storage cylinders.

A series of sudden and high intensity fluctuations can create oscillations in the system. Constant fluctuations of low intensity quite frequently appear in the delivery line. Disturbances created by aerodynamic wake after the impeller blades, flow over volute tongue, flow over diffuser blades, uneven angular velocity of the rotor (incase of gas and air pumping) are a few instances of creation of disturbances. If the frequency of the disturbance does not coincide with the frequency of the system, the amplitude of such oscillations give a very low effect on performance even when the pump works near maximum head point (K) and pump work under stable condition. If the disturbance frequency is very nearly or exactly equal to system frequency, resonance effect is created and the pump will work in the unstable operation even when pump is working far away (point A) from maximum head point K.

The frequency of operation of the system depends on dimensions of the delivery pipe and does not depend upon the speed of rotation.

A reduction in accumulator energy increases the frequency of oscillation which in turn increases the amplitude of pressure fluctuation. The energy waves developed due to pressure fluctuations, either direct and indirect, combined together, which can create high intensity shock waves. However, increase in frequency of the amplitude of fluctuations increases a self breaking effect.

A change in quantity of flow is created by the change in circulation, provided sufficient time is available. When total, combined self oscillating frequency of the system is reduced or in other words, when oscillating frequency increases, circulation time almost reaches the condition for one cycle of oscillation. This almost stops further increase in the amplitude of the oscillation, with the result; energy is dissipated under low accumulator energy. Circulation is inversely proportional to speed or tangential velocity of the blade. As a result, any pump working in the unstable region, when speed is reduced up to certain limit will work in stable condition at all flow rates. The limited speed will be higher where system conditions are low.

Boiler feed pumps, compressors working under high speed has a very little time to adjust, will undergo unstable operation even when accumulator energy is small. Stable operation in boiler feed pumps can be obtained by having continuous raising characteristics from full open to full close *i.e.*, without any drooping down characteristics at partial flow. Experimental investigation shows that reduction in number of impeller blades and low outlet blade angle β_2 will provide a stable raising pump $(H - Q)$ curve. Better results are also found when the impeller blades are extended into impeller eye at suction *i.e.*, increasing length of the blade at suction side. High specific speed pumps have raising characteristics.

Diverting part of fluid from delivery line to suction line is also one of the reasons for unstable operation for pumps.

5 Regulation

The process of changing the characteristics of system and that of pump to meet the output demand is called regulation of pumps. By this, the operating point of (H_{SY}–Q) and (H_p–Q) curves will meet the required output head and discharge.

Regulation is done by several methods namely (1) flow regulation by valve control system; (2) by transfer line control system; (3) by speed control of the pump; (4) by diameter control of the impeller of the pump; (5) by impeller and diffuser blade adjustments and (6) by changing the static height in suction and delivery tank.

5.1 REGULATION BY DELIVERY VALVE CONTROL

This method is simple, but with heavy hydraulic losses in the control valve. Regulation is done by a regulating valve fitted at the delivery pipe. Figure 5.1 shows the performance by delivery valve regulation process. In Fig 5.1, A is the operating point when the delivery control valve is at full open condition. Points A_{x1}, A_{x2}, A_{x3} are the operating points when control

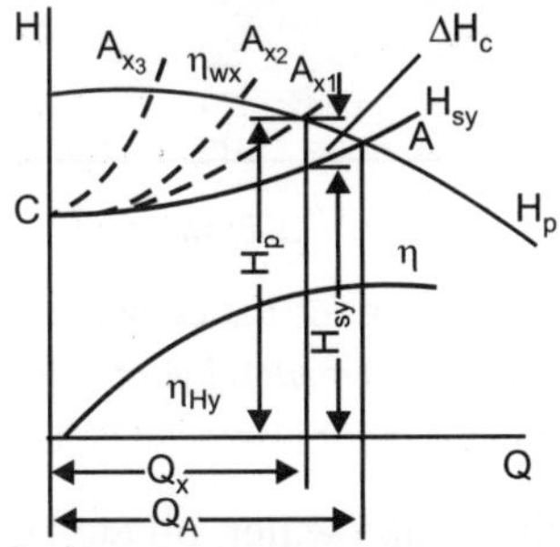

Fig 5.1: Regulation by delivery control valve

valve is gradually closed. Dotted lines corresponds to the (H_{SY}–Q) curves for different opening position of the suction valve. By adjusting the valve, either closing or opening, the (H_{SY}–Q) curve changes (curves A_{x1}, A_{x2}, A_{x3}). Each curve is for one position of gate valve opening.

Due to the adjustment in valve, the system resistance is changed by which, the operating point moves either upwards or downwards along the (H_p–Q) curve. Hydraulic losses are higher due to obstruction created by the valve and pump efficiency reduces considerably. However, due to simplicity, this regulation is widely used. Flow regulation by suction valve, although possible, but, this method is not carried out, because of low total head operation, which depends upon the delivery valve opening position. One such curve for one delivery valve opening position is shown in Fig. 5.2. The delivery valve controlled H-Q curve is also given in this figure for comparison purposes.

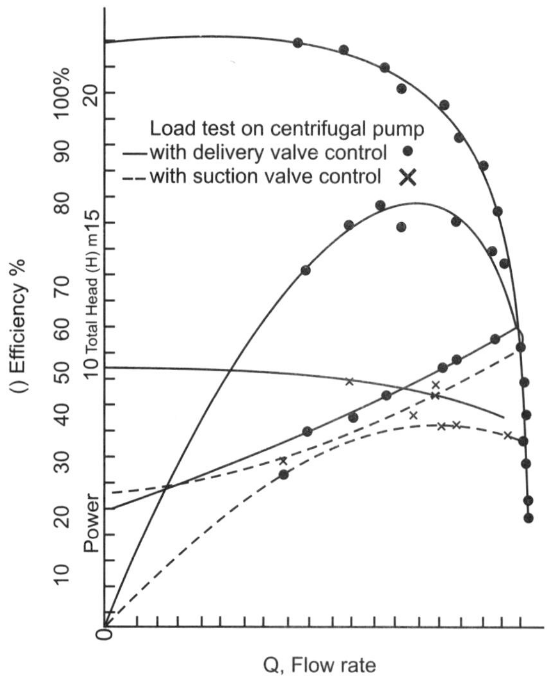

Fig. 5.2: Performance of pump

(a) Operation of delivery valve with full opening of suction valve

(b) By operation of suction valve for one position of delivery valve opening

For incompressible fluids like water, oil etc., due cavitations problem, suction valve controlled regulation is not used, since regulation of suction valve increase total suction head. However, for compressible fluid, suction valve control is carried out since density increases due to fall in pressure, which reduces the flow quantity. Moreover, unstable condition area moves towards left hand side of *H–Q* curve, more for air and less for gas. Figure 5.2 illustrates the graph for valve regulation by suction control valve and by delivery valve control.

5.1.1 Regulation by Transfer Line Method

In low specific speed pumps, the power of the pump reduces when control valve is gradually closed. Maximum power occurs when control valve is at full opened condition. But in axial flow pumps, power of the pump increases when flow rate is reduced or the control valve is gradually closed. In such installations a transfer line is always provided. The flow from the transfer line is sent back to the suction sump. When the control valve in the main line is gradually closed, part of the flow from the main line is transferred to transfer line. But pump supplies at constant flow rate, thereby power is kept constant and below overloading capacity.

This method is adopted in axial flow pumps in general for all pumps, where *N–Q* curve has raising characteristics, when flow (*Q*) reduces. In order to have minimum power at the time of starting, centrifugal pumps are always started with delivery gate valve in fully closed position while axial flow pumps are started with delivery gate valve in fully opened position. So a transfer line *i.e.*, a parallel pipe is connected to the main supply line. This transfer line discharges the pumping fluid back to the suction tank. While starting, the transfer line valve is kept open and main line valve is fully or partly closed. After starting, the main line valve is gradually opened and transfer line valve is gradually closed until required quantity in the main line is attained. Figure 5.3 illustrates the regulation by transfer line.

In order to maintain a steady flow, especially at low flow rate, recirculation method of regulation is often followed instead of regulating by gate valve. A part of the main flow from the delivery line is brought back to suction reservoir in order to maintain constant level. The suction reservoir is also constructed with greater depth. Referring to Fig. 5.4 the flow will be safe so long as the level in the reservoir is above the line 'a – a' *i.e.*, above the entrance of suction pipe. For the safe operation of pump without cavitation, the water level in the suction reservoir should be maintained at 'h_2'. If the liquid level falls below the value 'h_1', the recirculation line is automatically activated and the operating curve shifts from curve R to R + r. Out of total quantity 'Q_H' a quantity Q_C passes through the recirculation line and the balance through the main line (line r). The recirculation quantity maintains constant level of water in the sump and pump also operates safely. Recirculation line operates until the liquid level in the reservoir raises to 'h_2'. At this level recirculation line is closed. Between the point h_1 and h_2 pump operation is safe.

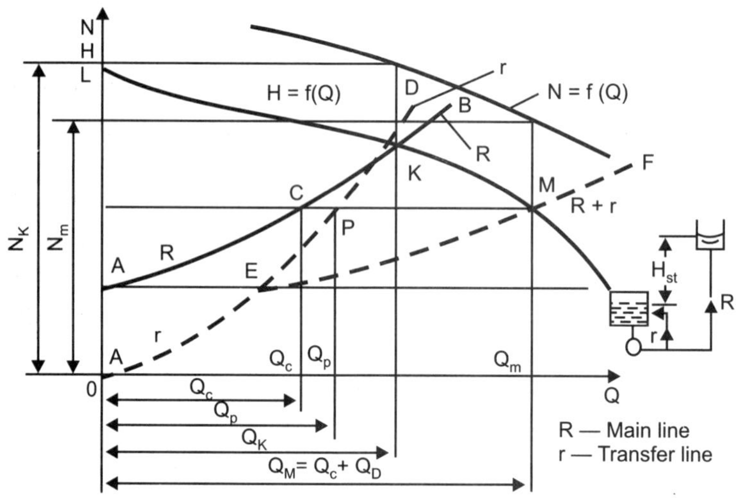

Fig. 5.3: Regulation by transfer line (r) OEPD-Transfer line, ACKB-main line®, OEMF-combined line (R + r)

In Fig. 5.3, parabola AB is the main line characteristics and parabola OD the characteristics of transfer line. OEF is the combined system characteristics (AB + OD) with transfer line opened. Parabola, LKM is the characteristics of pump. Under transfer line (OD) in closed condition, the operating point is $K.Q_k$ is the quantity through the main line AB. When transfer line is opened, operating point is $M.Q_c$ is the quantity passing through the main line and Q_p is the quantity passing through transfer line. Total quantity $Q_M = Q_C + Q_P$. The power N_M consumed is less than power N_K.

In Fig. 5.5, most commonly adopted circulation pump in ships is illustrated. For light weight provision, the outlet bend is provided with flow ribs and the inlet is provided with a main circular cylinder suction entry along with a side pipe entry. Due to the flow from the side entrance, unequal pressure and velocity distribution prevails at the inlet, which induces cavitation, noise, vibration. In order to improve the performance, inlet vanes are provided.

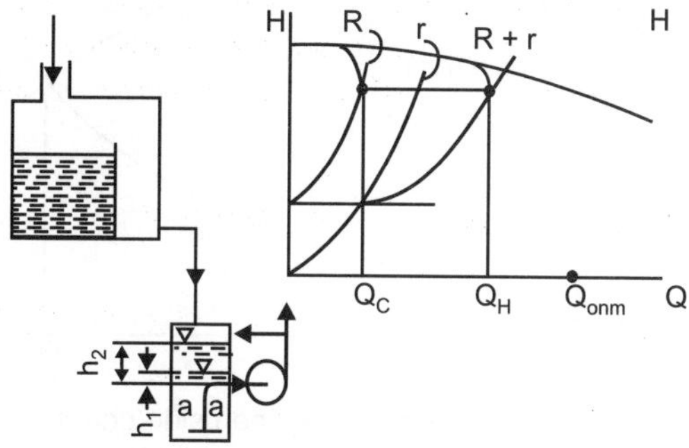

Fig. 5.4: Pump characteristics during constant recirculation through transfer line

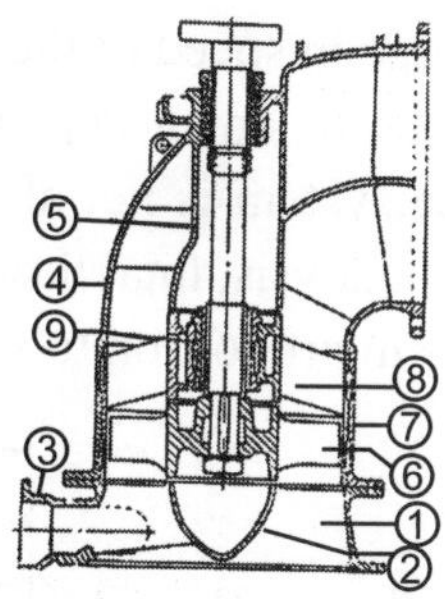

Fig. 5.5: Axial flow pump with suction control (side channel (3) flow controlled process)

Referring to Fig. 5.6 curve R is the *H–Q* curve of the axial flow pump, when flow is through the main entrance, this curve is less steeper. $H - Q_a$ curve is the performance of the same pump, when flow is through the side entrance. This curve R_1 and R_2 are more steeper due to heavy secondary losses at inlet. Experimental investigation shows that when the speed is reduced from $n = 100\%$ to $n = 60\%$ severe cavitation prevails, when flow takes place through side channel. The flow Q_a at $n = 60\%$ is found to be 15 to 20% of normal flow Q_H. Steady flow prevails up to the suction reservoir height up to 'h_{s_1}' where the inflow to the reservoir is equal to outflow from the pump. Steady flow exists during the height h_{s_1} to h_{s_2}. The flow becomes unsteady if the height in the suction reservoir falls below h_{s_1}.

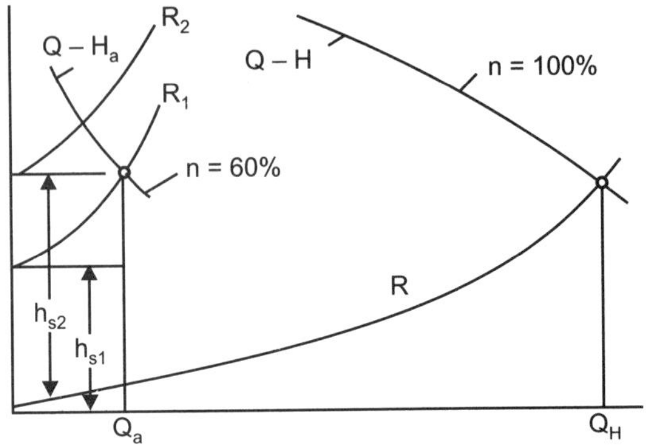

Fig. 5.6: Axial flow pump performance when side channel emergency line is in operation under speed change

5.2 REGULATION BY SPEED CONTROL

Already known that the change in speed of the pump, moves the $(H_p - Q)$ curve of the pump parallel up or down depending upon the increase or decrease in speed. Since the system is not altered, The $(H_{SY} - Q)$ curve remains same. As per model analysis, total head increases as square of the speed and quantity increases in proportion to speed.

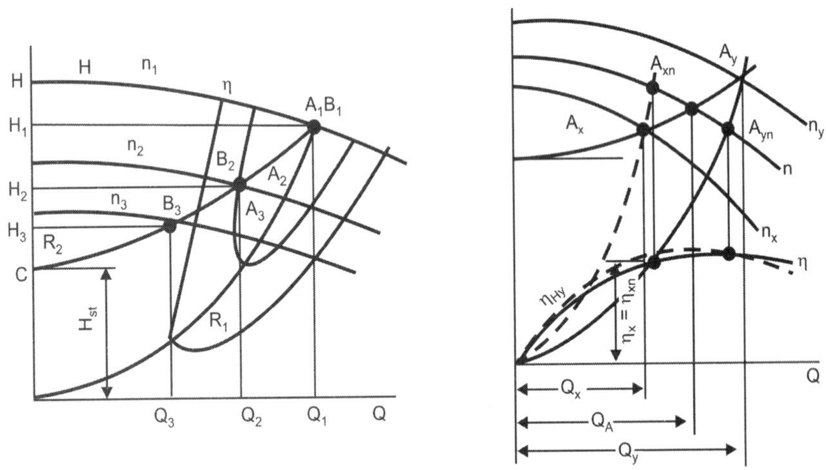

Fig. 5.7: Regulation by changing speed

In Fig. 5.7 and 5.8 $(H_p - Q)$ curves for three different speeds are drawn, over which the $(H_{SY} - Q)$ curve C, B_3, B_2, B_1 is overlapped. The operating points are B_3, B_2 and B_1 on each speed curve. It can be seen that although by model analysis points are obtained, efficiencies are not equal. At point 'B_1' efficiency is maximum. But efficiencies at other two point B_2, and B_3 are less than maximum efficiency.

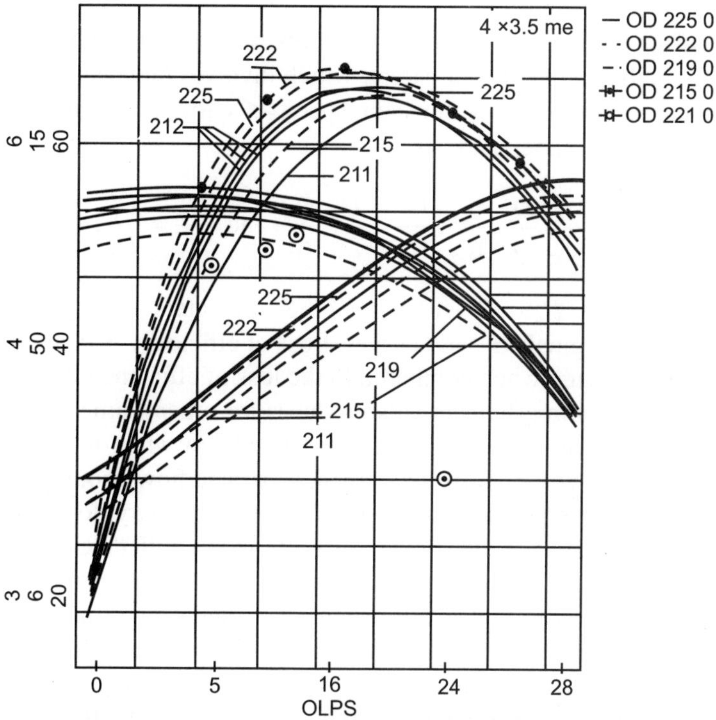

Fig. 5.8: Effect of changing diameter on performance

The $(H_{SY} - Q)$ curve must be altered in order to make the pump to work at maximum efficiency at all speeds. However, the loss in efficiency due to speed change is negligible when compared to other methods *i.e.*, regulation by delivery valve control method, as well as changing the impeller diameter keeping speed constant. This method of regulation is economical. There is no limit, provided prime mover capacity is sufficient to run the pump at such speed. Danger of cavitation is avoided completely. This method is adopted only for single stage units. For pumps in series or parallel, valve control method is always carried out keeping pump speed constant for all pumps.

5.3 REGULATION BY DIAMETER CONTROL (OUTER DIAMETER TRIMMING TO MEET THE DESIRED HEAD AND DISCHARGE)

Instead of change in speed, which requires a prime mover with speed changing facilities to a greater extent, which involves huge cost, a change in diameter of impeller at constant speed is carried out in chemical industries. As per model analysis,

$$\frac{Q_1}{Q_2} = \frac{D_1^3}{D_2^3} \cdot \frac{\eta_{v1}}{\eta_{v2}}, \frac{H_1}{H_2} = \frac{D_1^2}{D_2^2} \cdot \frac{\eta_{h_1}}{\eta_{h_2}} \text{ and } \frac{N_1}{N_2} = \frac{D_1^5}{D_2^5} \cdot \frac{\eta_{m2}}{\eta_{m1}}$$

Here also the $(H_p - Q)$ curve for other diameters, changes parallel with respect to original $(H_p - Q)$ curves. Loss of efficiency in changing the diameter is very high when compared to the loss of efficiency in changing the speed. Maximum efficiency also reduces for other diameters. This is due to the fact that flow at outlet is not streamlined flow when outer diameter is trimmed. There will be fully developed, separated flow at outlet edge of the impeller followed by an aerodynamic wake. Secondary flow increases considerably. Because of this, flow in diffuser and in outlet element is also a fully developed separated flow. As a result, hydraulic efficiency reduces to a greater extent; volumetric efficiency also reduces to a smaller extent. Mechanical efficiency very slightly decreases due to reduced disc friction loss. Overall efficiency reduces to a greater extent. However, in some impellers when impeller diameter is reduced very slightly, the efficiency improves slightly due to improved, better, flow pattern at the outlet of impeller at the passage between impeller, and diffuser, at the diffuser and at outlet elements. Figures 5.9 and 5.10 show the performance of pump with different diameters but with the same speed. The system curve is also shown in the same graph. Comparing the curves of Figs. 5.7 to 5.10, it is evident that efficiency drop in changing diameter is high when compared to the efficiency drop in changing speed. In Fig. 5.12, given the experimental results giving the limit up to which the diameter of impeller can safely be reduced to get better performance *i.e.*, without sacrifice in efficiency.

If the same impeller with different trimmed outer diameters is tested in the same spiral casing or diffuser, it is found, that η_{max}, for each trimmed diameter lies on the corresponding point, such that

$$\frac{Q_1}{Q_2} \approx \left(\frac{D_1}{D_2}\right)^3 \text{ and } \frac{H_1}{H_2} \approx \left(\frac{D_1}{D_2}\right)^2$$

The change of Q in proportion to D^3 is due to the change of C_{m_2} and the flow area 'A_2' perpendicular to C_{m_2} in direct proportion to the impeller diameter.

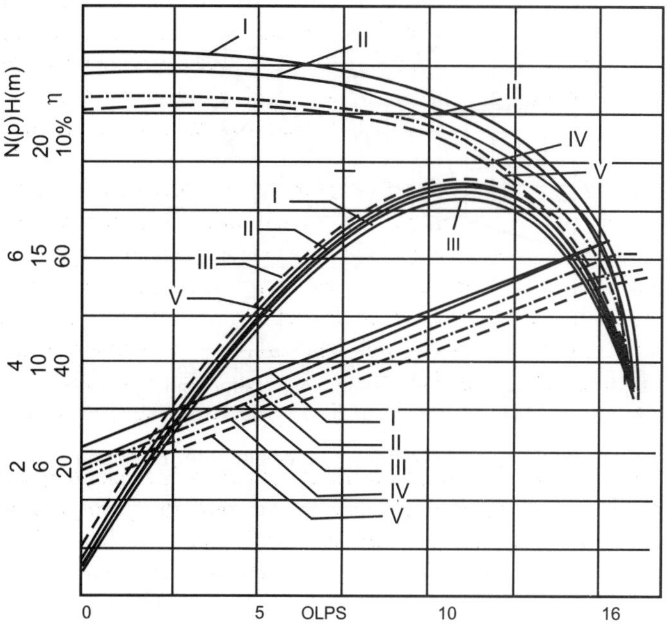

Fig. 5.9: Effect of changing diameter on performance

Referring to Fig. 5.12, the H–Q curve for the original diameter is curve AB. The required values of H and Q are given by point 'C'. A line is drawn from the origin 'O' to pass through 'C' and then to meet the original H–Q curve AB at pointy 'D'. This indicates that

$$\left(\frac{D_{2c}}{D_{2d}}\right)^2 \approx \left(\frac{Q_c}{Q_d}\right) \approx \left(\frac{H_c}{H_d}\right)$$

This method has been suggested by Berjourn of France and later accepted by other countries.

Although theoretically $H \propto D^2$ and $Q \propto D^3$ practically, the outer diameter 'D_2' must be trimmed to a value slightly higher than the value calculated as per the law $H \propto D^2$, in order to compensate for the quantity which follows $Q \propto D^3$. However, it is found that, this correction is not necessary if the law suggested by Berjourn is applied.

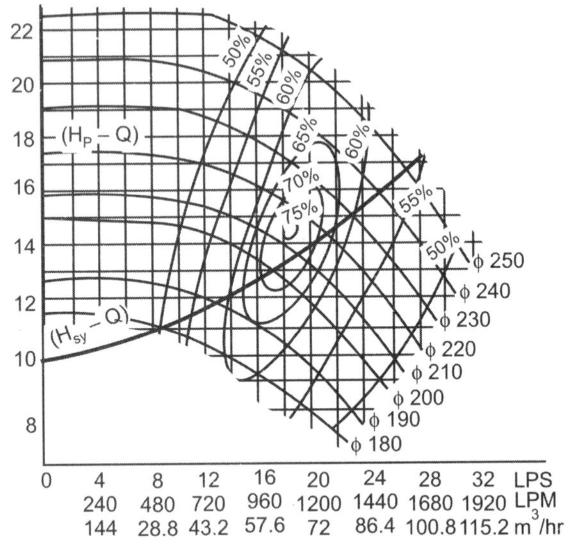

Fig. 5.10: (H-Q) curve for different outer diameter D_2 at one speed 'n'

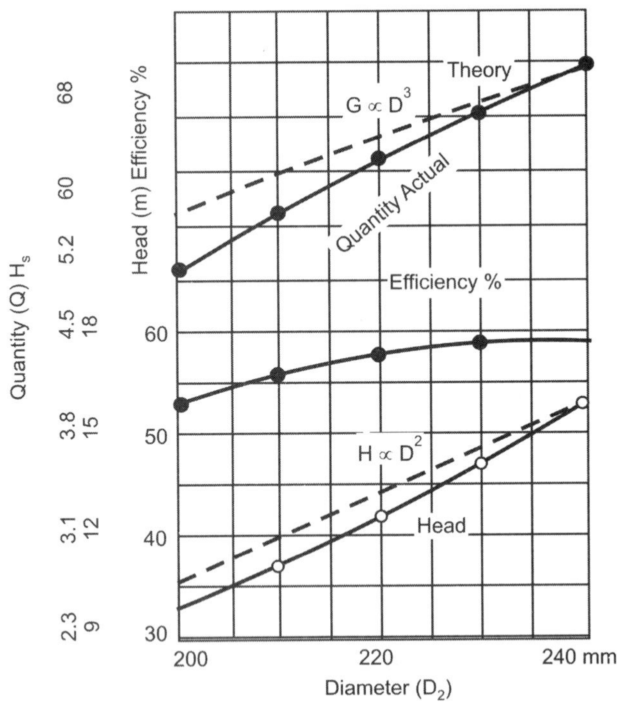

Fig 5.11: Effect of impeller outer diameters (D_2) reduction on pump performance

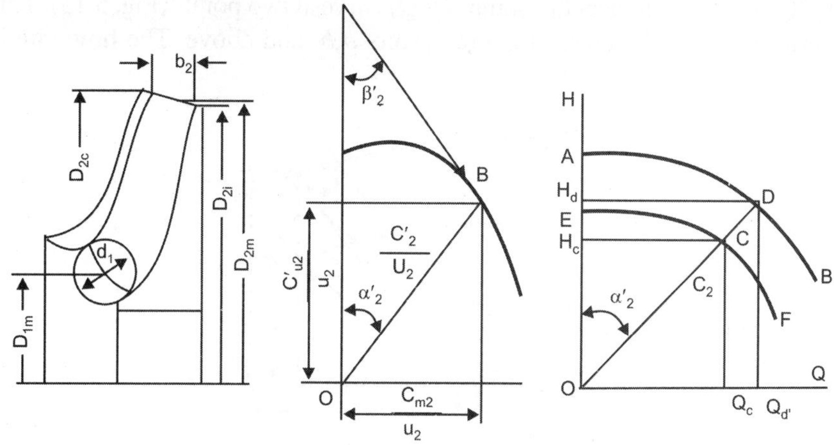

Fig. 5.12: For the calculation of probable diameter reduction

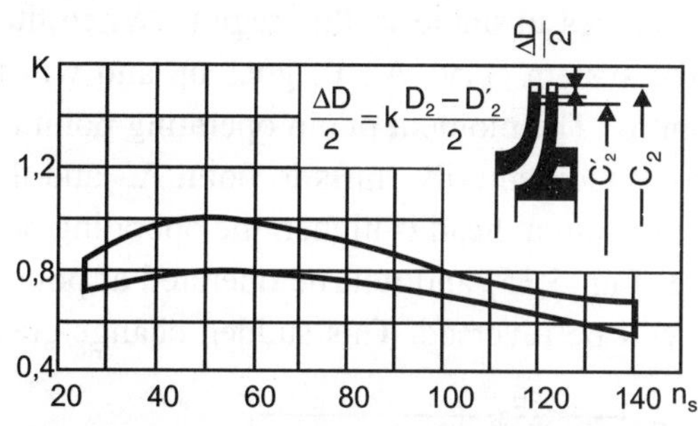

Fig. 5.13: Experimental coefficient k for turning impeller diameter with respect to η_s as per "RUTSCHI"II

5.4 REGULATION BY STATIC HEAD CHANGE

5.4.1 Static Head Fluctuation in Delivery Reservoir

In Figs. 5.14 and 5.15, pump (*H–Q*) curve with a drooping down characteristics regulation by static head change is shown. Water level in the suction tank is taken as (origin, O – O, *i.e.,* X-axis, Q-axis) reference line. Pump 1 supplies water through the delivery pipe 2 to the reservoir 3. Water level in the delivery reservoir corresponds to the static height H, H_1, H_2 and H_3 which is obtained by the meeting point of system curves AE, A_1E_2 and A_2E_2 and A_3E_3, with pump (*H–Q*) curve. Points B, B_1, B_2 and C are the corresponding operating points. Pump operation will be stable between heads H and H_2 *i.e.*, between points B and B_2. For further increase in height H_2 to H_3 by reducing the flow rate Q_{b2} to Q_{b3}, the system curve A_1E_1 raises parallel to X-axis and will meet the pump (H–Q) curve at two points (Fig. 5.15). This corresponds to the level in the reservoirs a_2b_2 and above. The flow rate is $0 < Q < Q_{b2}$ of the pump.

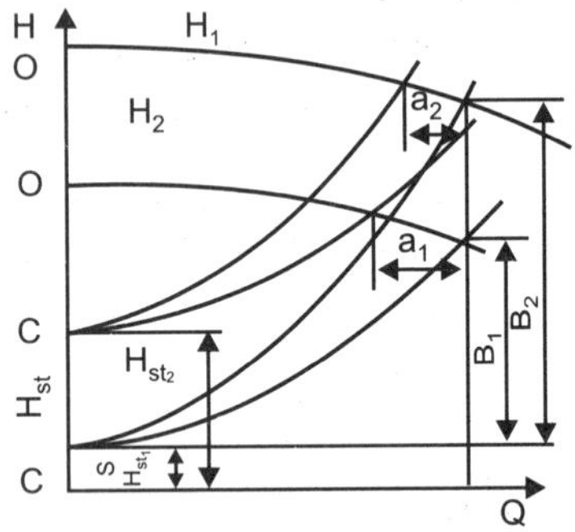

Fig. 5.14: Regulation by static head change

Operation becomes unstable in this region. When the pump head is raised further, the system curve A_2, E_2 goes up and will touch the H–Q curve at one point C. The moment pump operating point reaches point C, the operating point momentarily shifts to point A_2, and there will not be any flow. Since the system head is higher, the operating point further shift towards left side (Fig. 5.15) and will be operated at point 'A' (Fig 5.15), where the flow will be reversed. This sudden change creates huge noise

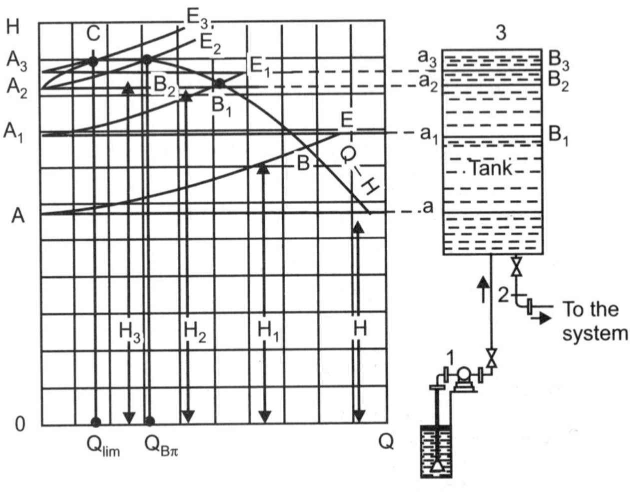

Fig. 5.15: Static head fluctuation in delivery reserve

vibration and water hammer which depends upon the flow rate and length of the pipe. Flow in positive direction will start only when the water level goes below a_1, b_1 which corresponds to the head H_1. Pump operation will always be unstable between H_2 and H_3 and above.

In order to avoid such unstable operation, pump (H–Q) curve must be gradual raising characteristics up to $Q = 0$, *i.e.*, without any drooping characteristics at low flow rate or other arise the curve $A_2 - C$ will be a raising curve as a continuous curve of C $B_2 - B_1$. This fluctuation of water level in delivery tank normally exists in systems, where condensers and accumulators are fitted in delivery line.

5.4.2 Static Head Change in Suction Reservoir

In Fig. 5.16(a), two different (H–Q) curves of pump are given $(H - Q)_a$ and $(H - Q)_b$. Curve CE is the system curve. Static H_{s2} corresponds to low water level in the suction reservoir, and H_{s1} corresponds to the high water level in the suction reservoir. Taking the low water level in the suction tank

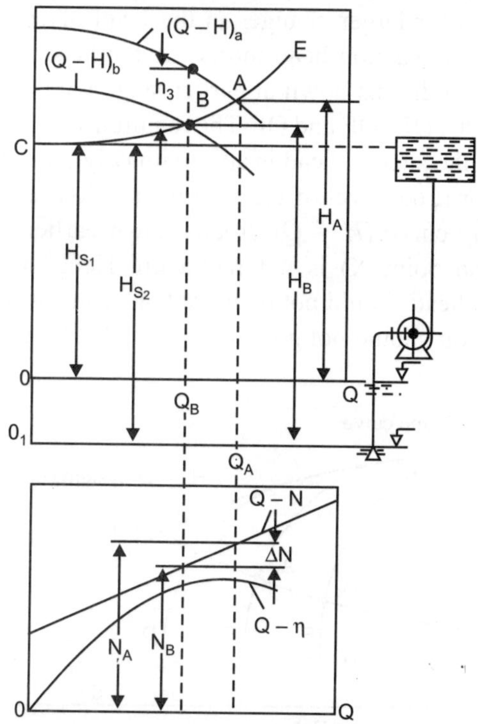

Fig. 5.16(a): Smaller static head change in suction reservoir

as the normal operating condition, the operating point is B with parameters Q_B, H_B N_B, corresponding to the point reaches the higher water level, the head will be H_A and parameters are Q_A, H_A, N_A, corresponding to the point A. The power increase $\Delta N = N_A - N_B$. Prime mover must be capable of taking this excessive power, if not; regulation is carried out by operating the control valve until Q_A reduces to Q'_A and N'_B. The difference in head $H_A - H_B = h_3$ is the loss created by the control valve. The control valve system, although simple but not efficient. The efficiency loss will be more. That's why when any system operates under larger and frequent suction or delivery reservoir level variations, system will be operated with another H–Q curve with larger drooping characteristics by this the change in Q, N, will be considerably smaller than in the lesser drooping characteristics of $(H - Q)$. However, best and efficiency method is always changing the speed of the pump. The same system is adopted when delivery tank fluctuations take place. This is applicable for smaller changes in the level of water in the suction sump.

In Fig. 5.16(b) the operating system $(H_{SY} - Q)$ with the pump curve $(H_P - Q)$ is given for larger changes in the level of water in the suction reservoir. When the suction head increases, the head discharge curve of the pump $(H_p - Q)$ droops down at earlier stage with respect to original, the curves OB, OC, OD, OE and OF. The required specification, marked as X, is met, when the suction head is approximately 2 m. When the suction head increases or if the level of water in the suction reservoir goes down further the pump curve $(H_p - Q)$ droops down earlier, and the required specification (the point X) is not met with. Hence for this pump the required suction head should not be more than 2 m and the level of water in suction reservoir should not go down below this level.

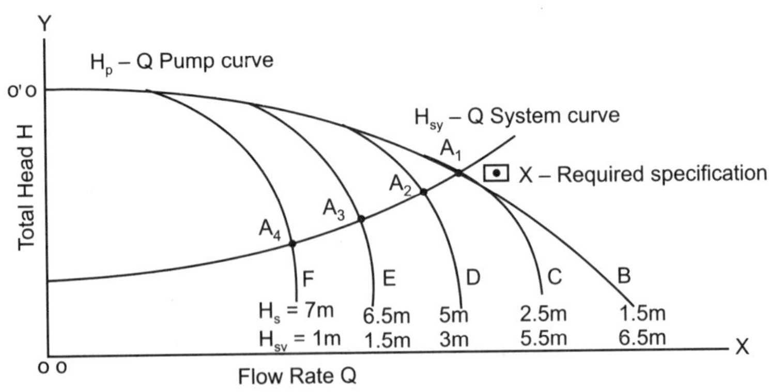

Fig. 5.16(b)

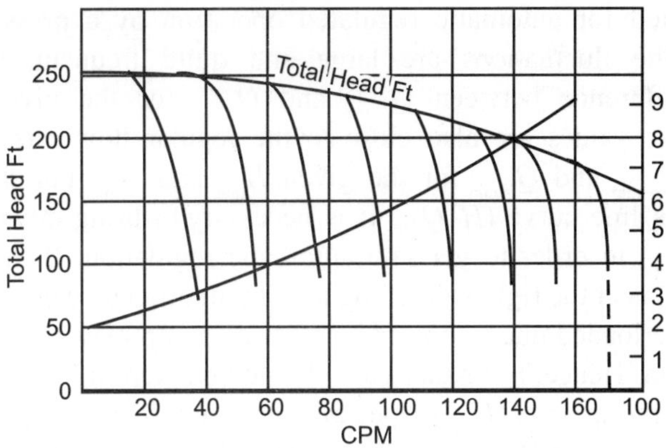

Fig. 5.16(b): Larger static head change in suction reservoir

5.5 REGULATION WITH TWO DIFFERENT (H-Q) CURVES OF THE PUMP

In Fig. 5.17, a system operation with two different (*H–Q*) curves, one with less drooping characteristics indicated by $(H\text{–}Q)_B$. Continuous line and another with higher drooping characteristics $(H\text{–}Q)_A$ indicated dotted line are shown. It is evident that drooping $(H\text{–}Q)_A$ curve (dotted line) is recommended

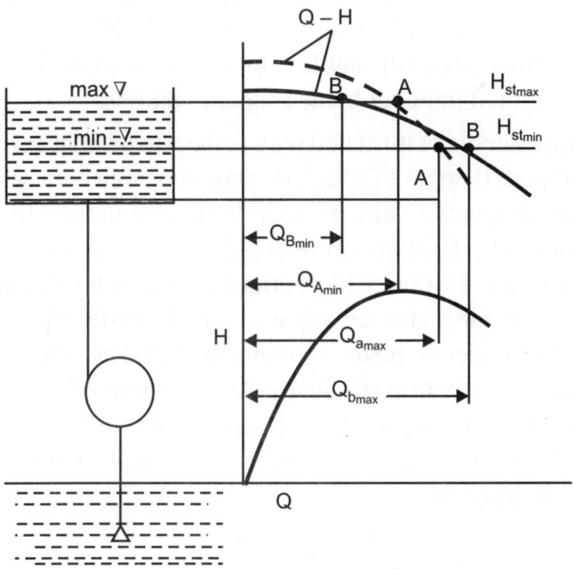

Fig. 5.17: Regulation by two different (H-Q) curves of the pump

and accepted for automatic regulated operation by a pressure control relay, if the fluctuations are large and quite frequent, because of smaller difference between $Q_{a\ max}$ and $Q_{a\ min}$ for the given H_{max} and H_{min}. These values are also close to the regular flow rate. But range between $Q_{b\ max}$ and $Q_{b\ min}$ for the given H_{max} and $H_{min.}$ For the curve B (continuous line curve $(H–Q)_B$. It is necessary to bring down the value $(H_{max} - H_{min})$ in order to get safe automatic regulation. If the operating point (H_{min}) is at the right side of the maximum efficiency point, the pump may be overloaded due to power increase or pump may be unsteady due to cavitation. Hence, necessary care should be taken in fixing the normal regular operating point (H_{min}) especially when automatic relay fitted. Pump automatically starts at (H_{min}) and automatically stops at (H_{max}). Normally this pressure difference will be 2 to 3 kg/cm^2.

5.6 REGULATION BY IMPELLER BLADES AND INLET GUIDE BLADE ROTATION

A small rotation of inlet guide blades before impeller blades, to some extent. Change the input energy to the impeller, which in turn changes the total head of the pump for the same quantity of flow. Efficiency almost remains same in this process. However, regulation can be done only to a very small range. Efficiency and head drop very much for further changes. Hence this method is not adopted in pumps. But the blade rotation is widely used with variable pitch adjustable impeller blade axial flow pumps. It is not possible in fixed impeller blade propeller pumps.

When impeller blade is rotated with respect to its own axis, the angle of attack is changed. If angle of attack is increased, area between two blades, axial velocity, tangential velocity, total head, circulation, lift, quantity of flow increases. The geometrical average velocity w absolute and blade angles also change. In Fig. 5.18 continuous line is for the original value and dotted line is for increased angle of attack. Reducing angle of attack by rotating the impeller blade in opposite direction reduces all values mentioned above including efficiency. At maximum efficiency, total head increases slightly because total head depends up on the curvature of the profile. As already explained in Chapter 10, it was shown that hydraulic efficiency η_h is equal to

$$\eta_h = 1 - \frac{w_\infty}{u} \frac{\sin\lambda}{\sin(\beta_\infty + \lambda)}$$

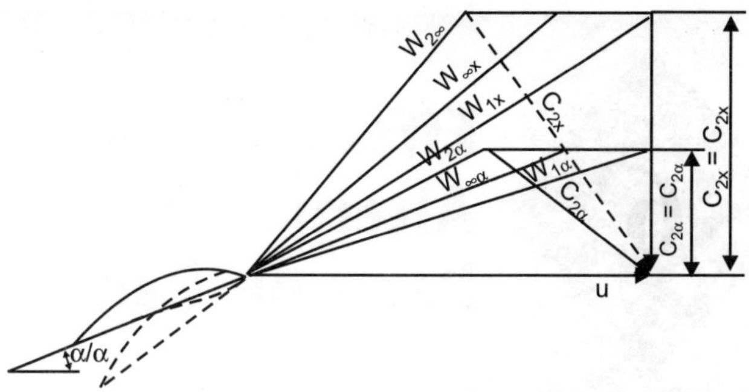

Fig 5.18: Regulation by impeller blade pitch control

Relative velocity *w* reduces only a little for a larger value of reduction in angle of attack. Value sin $(\beta + \lambda)$ reduces to a greater extent, with result, hydraulic efficiency drops when angle of attack is reduced or when impeller blade is rotated in the opposite direction $(-\Delta\beta_{\infty})$. Up to a reduction of $\beta_{\infty} = 23$ to 25, efficiency drop is only very little. Further reductions in angle of attack reduce efficiency significantly.

6 Effect of Minor Alternations in Pump Parts on Performance

6.1 EFFECT OF VOLUTE MACHINING

In some of the pumps, the volute casing is made in two halves, with a joint at the center, positioned with bolts, nuts and taper pins. Gaskets are used at joints for the prevention of leakage. Usually, these two joint surfaces are machined in a lathe or in vertical grinding machine. If these surfaces are machined more than the required size, the volute area will be reduced, which will result in lowering the efficiency and the optimum efficiency point will move towards lower capacity range. In order to avoid this, it is suggested to adopt rough grinding of the surfaces of the two volutes, which ensures: (1) Considerable reduction in machining time and cost due to minimum material removal and quick processing. (2) Positive leak proof joint with gasket because of rough surface and (3) Higher efficiency and more quantity due to higher flow area in the volute casing.

When volute area is larger than the designed value, maximum efficiency increases by about 7 to 10% and the optimum efficiency point may or may not shift to higher flow rate condition. This effect continues up to 10% increase in area. If area is further increased, there is no effect on performance. If the volute area is smaller than the designed value, maximum efficiency reduces definitely and the maximum efficiency point shifts to lower discharge position. Figure 6.2 shows the change in performance due to change in volute area. For example, if the gasket thickness is increased between two halves of the volute, the volute area increases and hence the performance of the pump may improve. In order to reduce the flow rate in pump, without any reduction in head, and loss of efficiency, a tapered inserts are introduced at the volute tongue portion.

(Fig. 6.1) By this more drooping down (H–Q) characteristics of pump is attained. Tapered inserts are mostly adopted where external control of flow should not be carried out.

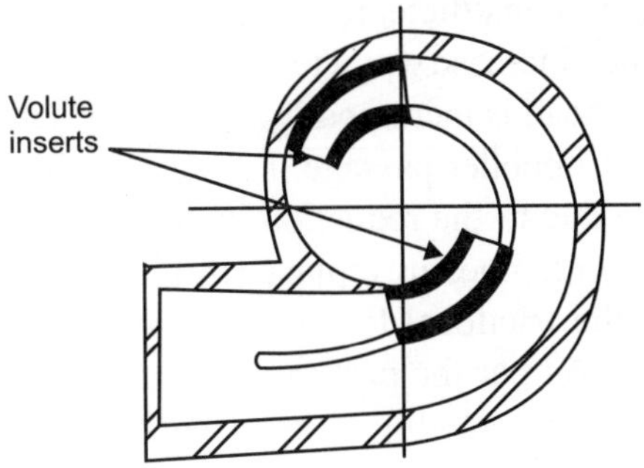

Fig. 6.1: Volute inserts convert high capacity pump to low capacity

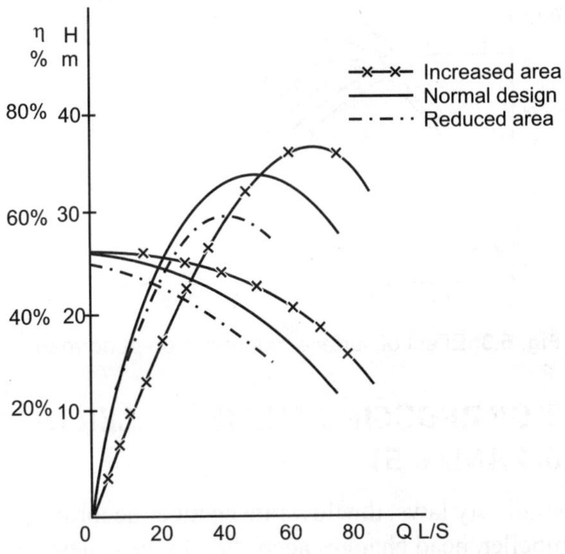

Fig. 6.2: Effect of volute area change on performance

6.2 EFFECT OF SURFACE ROUGHNESS OF IMPELLER VANE

Test was conducted on three types of impeller vanes (1) Impeller inner surface was coated with sand grain size of 0.5 mm, (2) Machined and polished vane surface of the impeller, and (3) Impeller inner surface is coated with plastic material to ensure perfect smooth surface of the flow passage.

Due to reduced roughness of impeller passages the polished impeller surface gave an increase in efficiency by 2 to 3% whereas in plastic coated impeller the efficiency increase was 5 to 6%. There is no change in quantity of flow, instead total head is increased from 2% to 6% depending upon the surface roughness of impeller passage. Figure 5.3 gives the performance change of the pump due to the reduction in surface roughness. The same effect prevails, when G.I. pipes are replaced by plastic pipes. Due to reduced surface roughness, the frictional losses are reduced, correspondingly the total head decreases. As per the performance graph, the quantity of flow is increased.

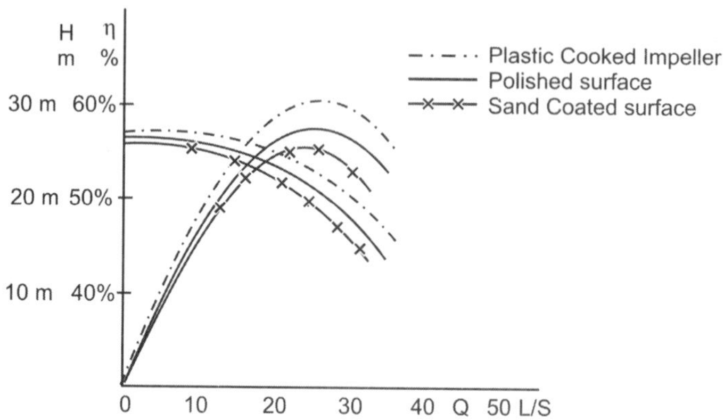

Fig. 6.3: Effect of surface roughness on performance

6.3 EFFECT OF REDUCING THE IMPELLER DIAMETER (Figs. 6.4 AND 6.5)

Although by similarity laws, the flow rate changes according to the cube of diameter of impeller, head changes according to the square of the diameter and the power changes according to the eighth power of the diameter. In actual practice, it changes still more. This is due to the fact that in similarity laws, efficiency is assumed to be constant. In practice, efficiency is not constant when diameter is reduced for the reasons explained earlier. Experiments conducted on different types of pumps indicate the following equation forms $H = D^{1.92}$; $Q = D^{2.82}$.

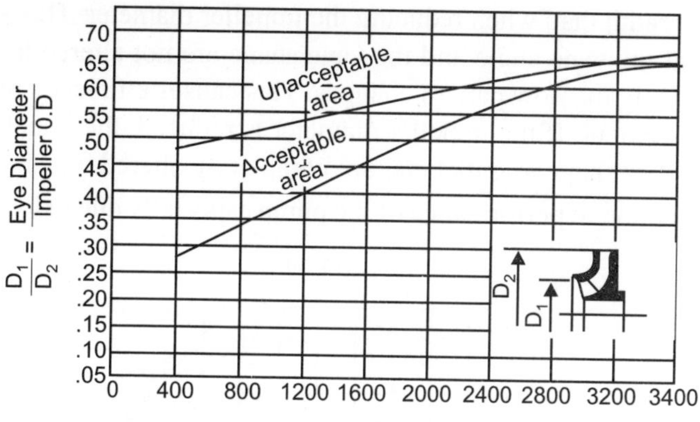

Fig. 6.4: Impeller eye diameter/outside diameter ratio

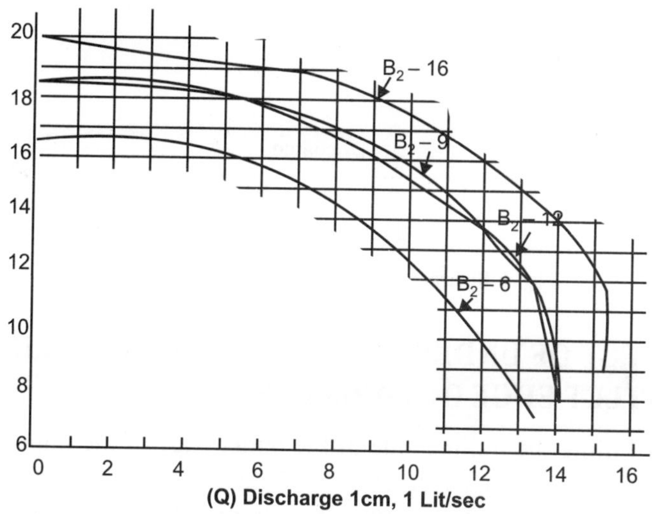

Fig 6.5: Head-discharge characteristics

6.4 EFFECT OF BREADTH (B) ON PERFORMANCE OF PUMP

Tests were conducted on the pump to determine the effect of blade breadth on the performance of a pump. Individual test results are given in Figs. 6.5 and 6.6. Efficiency increases by about 10% if breadth is properly selected. $(B_2/D_2) = 0.032$ at outlet gives the higher efficiency.

Experimental analysis shows that the rate of reduction in pump parameters such as head, discharge, efficiency and power is less when reducing breadth than when reducing the impeller diameter. This is due to the fact that the vane angle and the vane shape are not altered in breadth reduction. Only the flow pattern changes. This causes efficiency reduction to a smaller extent. If the impeller diameter is reduced, the impeller vane angle, impeller shape at outlet etc, are completely altered which causes a greater reduction in performance of the pump especially due to loss at exit.

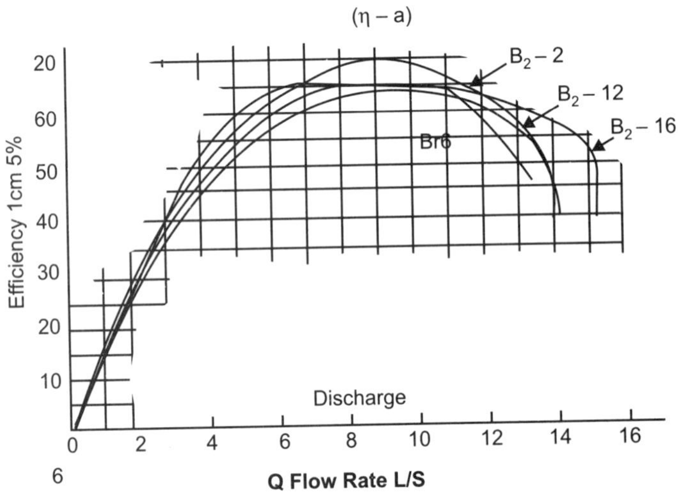

Fig. 6.6: Efficiency discharge characteristics

6.5 EFFECT OF UNDER FILLING AND OVER FILLING AT OUTLET EDGE OF IMPELLER

1. Filling the trailing side of the outlet edge of the impeller blade raises (*H-Q*) curve (Figs. 6.7 to 6.10.) to a certain extent. Efficiency may remain same or may increase. But normally decreases.
2. Filling the leading side of the outlet edge of the impeller blade drops (*H-Q*) curve. Efficiency also drops.
3. Instability in (*H-Q*) curve at low *Q* can be removed shifting the inlet edge of the impeller blade towards eye.
4. Thinning the inlet edge by filling both sides of the blade especially more on the leading side, less on the trailing side and rounding off the inlet edge reduces drooping nature at unstable region.

5. Increasing the eye diameter D_0 near inlet edge of the impeller blade and sometimes shifting the inlet edge inside improves NPSH.
6. Increasing the blade length at hub and decreasing the blade length at periphery of the inlet edge, in other words, increasing the inclination of inlet edge with respect to outer shroud improves NPSH.
7. Filling the volute tongue removes the unstable nature at low Q, but Q reduces at higher flow rate.
8. Increase in volute area may increase the overall efficiency but optimum efficiency shifts to higher Q. Reduction in volute area will reduce overall efficiency and the optimum --- shifts to lower Q.

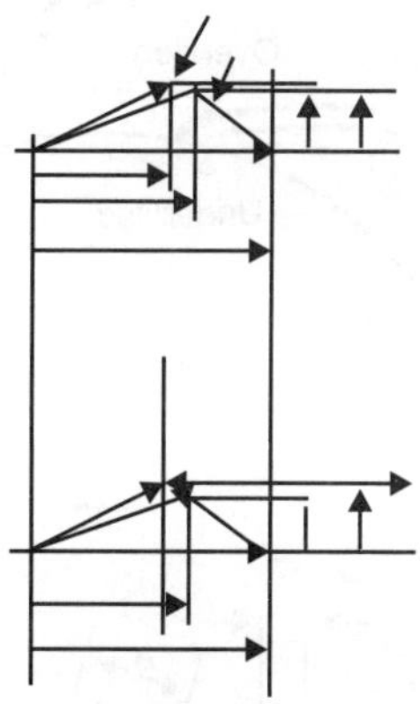

Fig. 6.7: Discharge velocity for under filied vanes

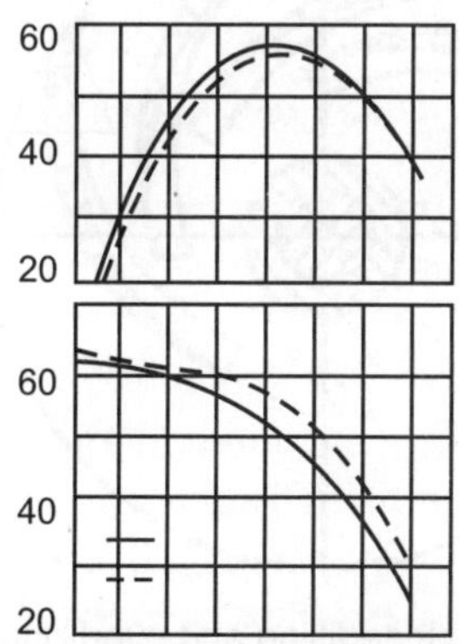

Fig. 6.8: Effect of tapering-off the trailing side of the impeller exit tips

Increase of total head without any change in rate of flow, is carried out by under filling the impeller outlet edge (Figs. 6.9 and 6.10). Outlet blade angle of the impeller slightly increases which in turn increases the total head. Such process is done, when the existing pump develops slightly lower head than required. An increase of 5 to 10% can be increased. Excess filling reduces efficiency and total head, due to high hydraulic losses created by the secondary flow and wake formation. Figures 6.9 and 6.10 show the place of under filling and the change in performance due to this. Correspondingly over filling, *i.e.,* filling at the leading side outlet

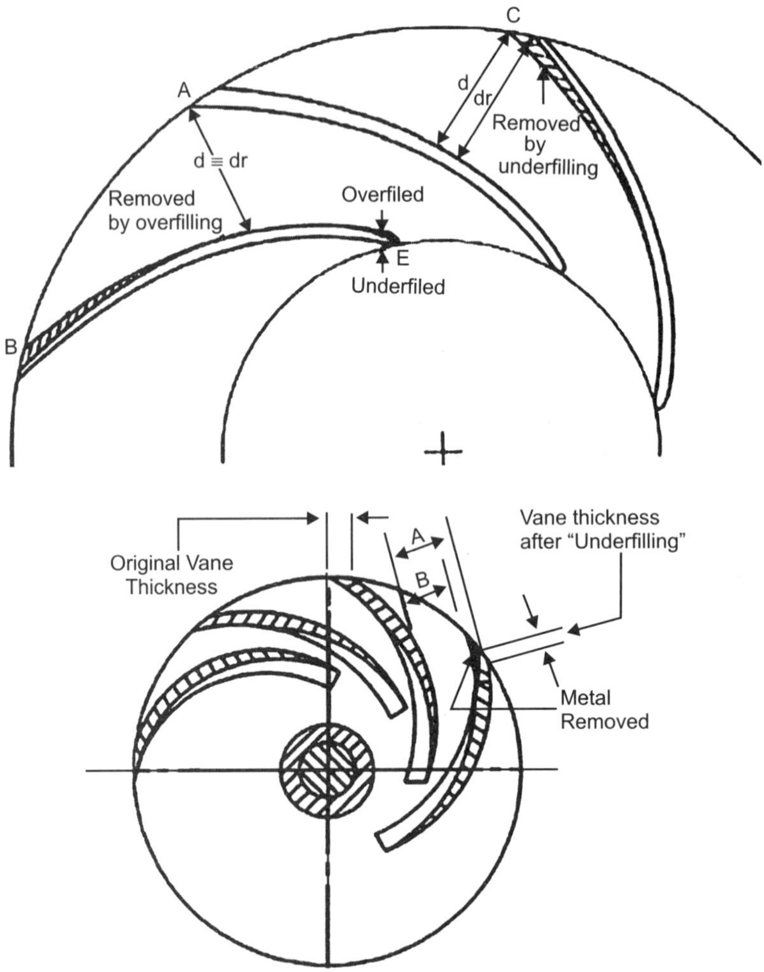

Fig. 6.9: Underfilling and overfilling of vane

edge may reduce total head but efficiency reduces considerably. Outlet edge of the impeller controls the total head and inlet edge of the impeller controls NPSH. Figure 6.9 shows the variation of *H–Q* curve due to the introduction of above mentioned process.

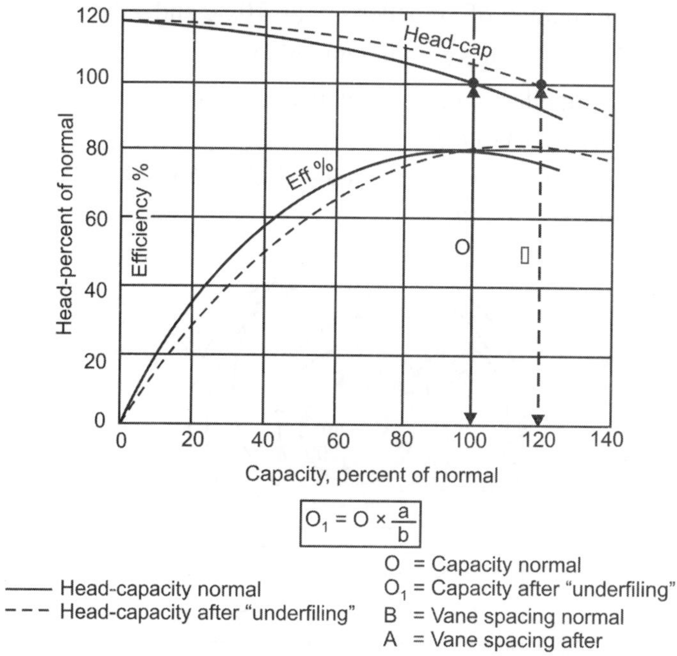

Fig. 6.10: Changing pump performance by impeller underfiling

6.6 EFFECT OF INACCURATE MANUFACTURING ON PERFORMANCE

19 impellers of a pump namely one with correct manufacturing and other 18 with defective manufacturing were tested. In order to find the effect of manufacturing in accuracies on pump performance the following variations were made:

1. Shift of outlet edge in circumferential direction:
   (a) 3 impellers with +ve angular shift $(+\Delta\rho_2)$ + 5°, + 10° and + 19°.
   (b) 3 impellers with –ve angular shift $(-\Delta\rho_2)$ +5°, 30 and + 13°, 20, + 20°, 20.

2. Shift of inlet edge in circumferential direction:
   (a) 3 impellers with +ve angluar $(+\Delta\rho_1)$ +4°, 50 and + 9°, 30, and + 20.
   (b) 3 impellers with –ve angular shift $(-\Delta\rho_1)$ along the direction of rotation of impeller – 5°, – 10.30 and – 20°.
3. Shift of inlet edge in radial direction
   (a) 3 impellers with +ve radial shift $(+\Delta r_1)$ $(r_1 + 2.8)$, $(r_1 + 4.3)$ and $(r_1 + 6$ mm$)$
   (b) 3 impellers with –ve radial shift $(-\Delta r_1)$ $(r_1 - 1.7)$, $(r_1 - 4)$ and $(r_1 - 6$ mm$)$.

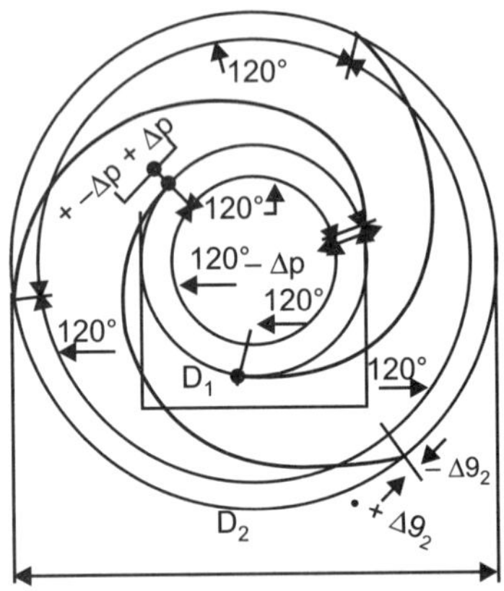

Fig. 6.11: Intentionally made workmanship defects

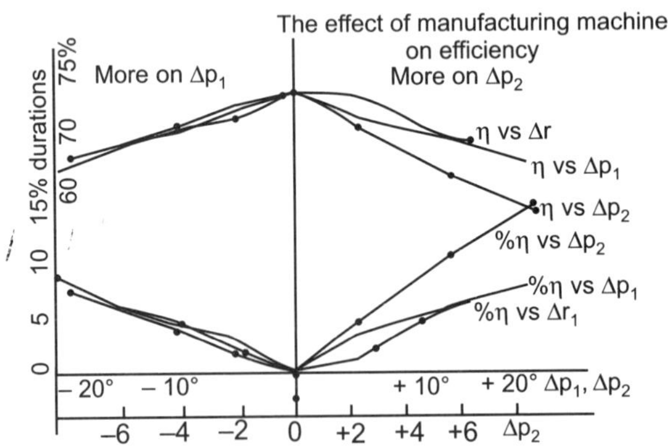

Fig. 6.12: Effect of manufacturing inaccuracies on efficiency

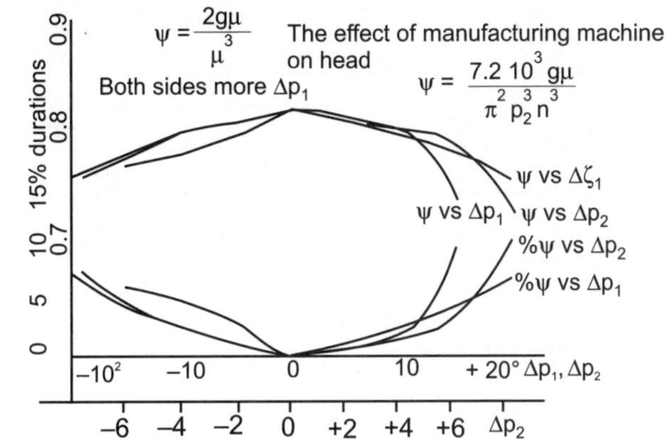

Fig. 6.13: Effect of manufacturing inaccuracies on head

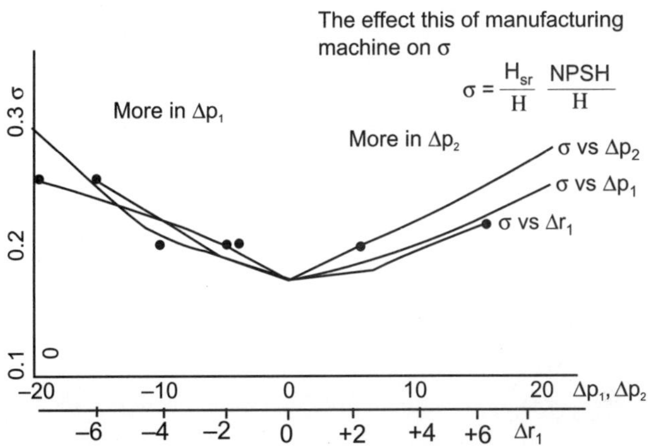

Fig. 6.14: Effect of manufacturing inaccuracies

Figure 6.15 shows the details of the above mentioned defects and Figs. 6.11 to 6.15 give consolidated test results of the variation in head, efficiency and cavitational constant (Thoma's constant σ (NPSH = σ*H*). It is found that head loss, efficiency loss are more in + P_2 *i.e.*, angular shift of outlet edge opposite to the impeller rotation. The percentage loss of head is almost 11% and the percentage of efficiency loss is 12%. σ changes from 0.15 to 0.28 almost twice. Hence it is recommended that impeller blades must be manufactured with proper care especially the outer edge.

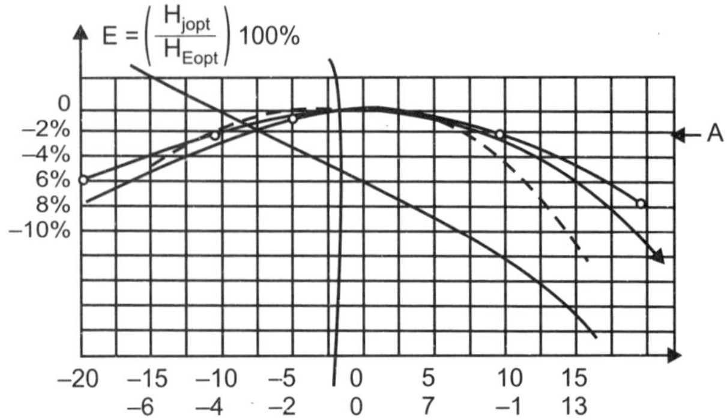

+ – Angular shift inlet edge O – Angular shift outlet edge X – Radial shift

Fig. 6.15: Variation of the delivery head in % depending upon the various inaccurate workmenship

7 Pump Maintenance

Failure of pumps, running in line may be gradual like backlash of liquid causing blistering of impeller or casing, or sudden, like shaft break. Both of these failures are not common in well operated and maintained pumps. Incorrect application like improper materials used in different parts of a pump can be easily traced; example of incorrect application is use of bronze fitted pump for pumping alkaline solution having a high pH value. Another problem is poor operating conditions and maintenance. Pumps wear out gradually, mostly due to corrosion of impeller vanes. Wrongly applied pumps must be replaced before too much trouble starts.

Hydraulic failures like, not delivering or low discharge of liquid, low delivery pressures are most common and very difficult to solve. Almost 85% of troubles in pumps occur at the suction side of the unit. Some of the causes and remedies are analyzed.

7.1 PUMP NOT DELIVERING LIQUID

Reason may be pump not primed *i.e.*, the suction pipe, impeller, volute casing and part of delivery pipe must be filled with the pumping liquid. Air vents kept at different places must be kept open, while filling, and must be closed after filling. Filling should be carried out until liquid comes out from the air vent *i.e.*, air must be completely removed.

7.2 PRIME MOVER SPEED IS LOW

The electric connection to the motor, if fitted as prime mover to the pump, must be checked. All three phases must be properly connected. Correct voltage must be applied to the motor. The governors must be checked for speed control in turbine driven units. Flow of steam in pipeline, through valves must be checked. Plugging in steam pipe and worn out valves may reduce the steam pressure and flow rate. Intake air supply, fuel supply, (injection nozzles, carburetor, magnets), speed control setting in governors, in engine drivers units, must be checked. Supply air pressure, near the air driven prime movers, connecting pipelines, hoses, even reduced pipe or hose size may cause reduction in the speed of the air driven unit (excessive use of air at some other places, open drain pipe etc.).

7.3 HIGH DELIVERY PRESSURE

Obstruction on clogging in suction and delivery pipelines, either due to gradual accumulation of foreign materials such as non dissolving solids or scale formation, valves in partly closed condition are the causes. If clogging is in suction pipeline, pump may develop noise and vibration due to cavitation and subsequent erosion of material. The total load developed may be lower than the normal valve. The suction gauge will read low pressure or high vacuum. The discharge will be lower than the normal value. If the discharge is also high along with delivery pressure, prime mover speed will be higher and should be reduced.

7.4 SUCTION LIFT IS TOO HIGH

High vacuum, *i.e.,* pressure below vapour pressure, before the inlet to pump, clogging in foot valve, suction pipe, suction valve, reduced suction pipe size. Suction lift is too high, air presence, as air pockets, suction pipe not sub merged sufficiently in suction tank are also the causes for no discharge. NPSH should be checked. Raising the suction chamber to a higher level or lowering the pump level, increasing the suction pipe size, cleaning the entire suction line are the remedies. For centrifugal pump a suction lift of 6.5 to 7.0 m is recommended.

7.5 REVERSED DIRECTION OF ROTATION

Check the electric connection for the electric motor and correct it. The direction of rotation should be as per the arrow marked on the casing.

7.6 QUANTITY DELIVERED IS LOW

Reversed direction of rotation, worn out wearing rings, air entry through stuffing box, damaged impeller, low level of liquid in suction chamber, low speed of rotation, total head is higher than the designed head, air entry into suction line, high suction lift (in high NPSH), suction pipe not submerged sufficiently, clogging in suction pipeline, foot valve and in suction valves either due to mud or foreign material or due to scale formation, damaged foot valve are main causes for low discharge. Air leaks can be checked by the following procedure: close the suction valve and fill up the suction pipeline with the liquid under pressure. The liquid leaks out of the air pockets. A pressure gauge can be connected to the suction line. If the pressure falls after some time, it indicates that there is leakage. The best method is to unscrew, all screwed joints, put leak proof paste on the thread and fit it again. The leathers or rubber washers used should be replaced with new washers. The stuffing box packing *i.e.,* asbestos material should be changed. The gland bolts are tightened such that the fluid comes out in drops from the stuffing box. For mechanical seal units, the gland washer should be checked and replaced if necessary for air tight. The stuffing box pressure should be slightly higher than atmospheric pressure (1.2 to 1.5 ata). Shaft sleeves, if worn out must be replaced. If there are lantern rings in stuffing box, for stuffing box cooling it should be cleaned either by pumping liquid or by external means. If pump produces noise and vibration, low NPSH may be the reason for which the manufacturer must be consulted. If the wearing rings are worn out, it must be replaced. Recommended wearing clearance must be maintained (always 0.15 to 0.25 mm). Undersize foot valve must be replaced, especially the one way opening leather washer for foot valve. The foot valve for the horizontal pumps, the bell mouth entry for vertical submerged pumps, must be placed sufficiently away from the suction chamber wall so that the liquid can enter uniformly to the suction pipe from all sides. The foot valve or bell mouth entry should be at least '1*d*' distance from the wall on all sides (*d*-the diameter of the pipe) and 1 to 1.5*d* from the bottom floor of the suction tank. The submergence must be a minimum of 1*d* to 1.5*d* depth from the free surface of the liquid in suction chamber. In case a number of pumps are operated from a common sump, the suction sump must be properly designed to ensure even distribution of liquid to all pumps, from the intake pipe of the sump. If viscosity or consistency of the liquid is high, the discharge and the total head will be low, power consumption will be higher than normal. The pump selection may be wrong or liquid might have been changed. If entry to the impeller from the suction pipe called "impeller eye" diameter is less than the suction pipe, the discharge will be low. This is due to improper matching of suction pipe and pump inlet diameter. Always impeller eye diameter must be equal or higher than the suction pipe. The size of suction pipe must be checked for proper value of NPSH and eye velocity, C_0. Sometimes faulty flow meters may also indicate low discharge. Faulty pressure and vacuum gauges may indicate incorrect pressure head at delivery and at suction lines. Periodical calibration is necessary.

7.7 POWER CONSUMPTION IS HIGH

If the discharge is too high, because of increased speed of pump, may be due to high voltage at the electric motor, improper setting of governor in steam turbine or in I.C. engines, power consumption is high. Voltage should be regulated by voltage regulators. Governor setting should be corrected by adjusting the speed setting screw in throttle line of I.C. engine, or in the steam stop valve to provide proper opening and for correct fuel supply. In case the speed is proper, but the discharge is high, the total pressure head will be low, but power consumption will be high. In such cases, partly close the delivery line valve only and not the suction valve. If the suction valve is operated, then NPSH will be different and pump may not deliver liquid. The shaft sleeve or housing sleeve in stuffing box might have been worn out, thereby concentricity between shaft and housing at the stuffing box is not existing. This in turn, will not rotate, the impeller, concentrically with the casing. Impeller wearing ring may be touching the casing wearing ring at the bottom (mostly) which increases the power consumption. Wearing rings and shaft and housing bushes must be changed and proper clearance between sleeves and wearing rings must be maintained. All these clearances must be properly lubricated by the pumping liquid. Occasionally worn not bearings misalignment of bearing's, tight bearings may cause shaft concentricity with housing. In case of such problems, noise, vibration and heating at the bearing housing can be heard. Bearing must be changed. Sometimes wrong lubricant, improper lubricant or no lubricant can also create such problems. Reversed direction of rotation can cause either lower or higher power consumption. In such cases, head and discharge will be very low. Improper selection of pump is also one of the reasons of higher power consumption. Pumps must be selected strictly according to the specific speed and liquid distorted impeller or casing suction and delivery piping of pump, worn out shaft, misalignment of pump, wearing ring, bushes, bearings consume excess power. If stuffing box packing is too tight, power consumption will be high. In such case change the packing and adjust the gland pressure to 1 to 1.5 ata. Mostly, vertical pumps are self lubricated, *i.e.*, lubricated by the pumping liquid. It does not contain impurities or suspended solids. In case of water pumps, pump shaft bushes and sleeves will be submerged in water and hence self lubricated. When water level goes down, bushes may not be lubricated, which is the cause of extra power consumption. Sometimes faulty ammeter, wattmeter or energy meter may also indicate excess power consumption.

7.8 PUMP VIBRATES EXCESSIVELY

Rough operation of pump, improper maintenance of pump, not following the instructions for periodical check up procedures, excessive wear in pump parts, foundation bolts, not properly tightened, are some of the causes for vibration of pump. Air leakages into the pump, not maintaining the recommended NPSH as given by the manufacturer are also the causes for vibration of pump. Insufficient submergence of suction pipe with foot valve or bell mouth make the air to enter, in the form of free vortex, into the pipe line. This can be seen directly by naked eye by standing by the side of the suction sump and viewing the water level. Screw type or thread type revolving vortices are formed due to air entry, which extends up to foot valve or bell mouth. Air leakages at stuffing box, in suction pipe lines through joints in valves especially through the gap between valve stem and valve gland and body are some of causes for air leakages. Higher suction lift, smaller suction pipe diameter, too many fittings in suction pipe, such as valves, bends, tee joints, clogging in suction pipe and in foot valve cause NPSH to be lower than the recommended value. This also gives noise and vibration due to cavitation and subsequent erosion especially at the impeller blade inlet edge. Bearings, not properly lubricated, presence of too much foreign materials such as sand, wood in general non dissolving solids, create misalignment of pump, which may be reason for vibration of pumps.

7.9 PUMP IMPELLER CRASHES WITH THE CASING, IMMEDIATELY AFTER STARTING

In case of single stage pumps if the coupling nut, or bearing collars or impeller nut are not properly tightened, the impeller rotates at the time of start but the impeller with shaft moves forward towards suction pipe and crashes (or jams) with the casing and stops rotating. In fact, after stopping it is not possible to rotate freely by hand. The same thing takes place in multi stage pumps, having in line impellers (not opposed impellers). This is due to the axial thrust developed when the pump runs (refer axial thrust section). Axial thrust increases, when the pump speed increases. The tendency of axial thrust is to push the impeller with or without shaft towards the suction side of the casing and to crash against casing, whatever may be clearance between impeller front (or suction) side shroud and suction side of the casing. When the pump stops the impeller does not return back to its original position but remains at the crashed position. The pump shaft cannot be rotated. A slight hit with a hammer at the suction side of the casing makes the impeller to go back to its original position and makes the pump shaft to rotate freely. The impeller nut, bearing nut, coupling nut must be checked and tightened by a torque wrench. The same reason applies to multi stage pump also. The pressure difference between inside and outside of the balancing devices in multi stage pumps must be equal to the pressure difference between the delivery of the last stage of impeller and suctions of the first impeller. If this is not maintained then, impellers will definitely jam or crash immediately after starting. To maintain this condition, a pipe is connected between the suction side of impeller and the outer side of the balancing disc. If this pipe is absent, crashing can take place. Also, if the connecting pipe diameter is not enough to maintain correct flow from balancing disc to the suction side of the impeller, in order to maintain equal pressures at both ends of the pipe (without loss in pipe), due to difference in pressure difference at the impeller and at the balancing disc, the pump shaft jams. In such cases the connecting pipe size should be as large as possible, that should be permitted to suit the pump construction.

7.10 DISCHARGE RAPIDLY REDUCES AFTER STARTING THE PUMP

This is what, is called losing of priming after starting the pump. This is due to leakage in suction line. All suction pipe joints, including suction flange connection with the pump must be checked. Suction flange gasket should be changed. Joints are pasted with leak proof pastes and then connected to each other. Stuffing box packing should be replaced, if air enters through stuffing box. Ensure always that, at least a drop of liquid per 10 seconds must leak out from the gland to atmosphere. Lantern ring must properly be put up in stuffing box at proper position. Cooling liquid must pass through the lantern ring freely. Proper submergence must be maintained, if not, either pump should be lowered down or water level in the suction chamber must be raised, so that air entry into the foot valve is stopped. Improper priming is also the cause of this effect.

7.11 OVER HEATING OF STUFFING BOX

This is a common cause of trouble. Packing materials may be old and very hard. Sometimes excess tightening of grand to the packing can also make the new packing to become hard. No cooling water supply to the lantern ring, even improper installation of packing may cause such troubles. Packing manufacturer should be consulted.

7.12 BEARINGS WEAR AND OVERHEAT

Improper lubricating oil, low oil level in oil case, dirt, moisture in oil, clogged or sealed oil cooler, not sufficient supply of cooling water to the bearing chambers, bearings too tight or oil seals fitted too close to the shaft, misalignment of bearings due to excess wear, bend in shaft, excess axial thrust exceeding the bearing capacity, wrong bearing installation are the main reasons for bearing vibration, overheating and subsequent noise.

7.13 PUMP DOES NOT START

Impeller and casing, and/or, bushes and sleeves of stuffing box, touching each other due to worn out of wearing rings, insufficient bearing capacity (thrust bearing or ball bearings) to take care of axial and radial thrusts, dirt or sand or corrosion growth or any solid particles sticking on, at the clearance between wearing ring and at stuffing box sleeve or bush, no lubrication of bearings thereby bearings overheat and seizes are some of the causes. If the pump was kept idle for a long time, it should be dismantled, overhauled, cleaned and reassembled, take care to change worn out parts, providing proper lubrication to bearings, proper cooling liquid supply to stuffing box, bearing chambers. Packing seal should never be over tightened, which is also one of the reasons. If the jamming of impeller with casing is due to axial thrust, impeller, shaft assembly should be pulled back slightly by the nut provided at the other end of the shaft. At the same time, proper clearance between impeller and casing especially in case of semi open impellers must be maintained; otherwise, efficiency will drastically reduce (by 15% to 20%). There are so many troubles in pump operation. Only a few are explained. However the remedies explained above, will mostly be same for most of the troubles. Periodical maintenance, servicing, following the manufacturers instructions will give a long smooth, trouble free operation of pumps. In the following pages, causes and remedies for pumps are given in short form.

7.14 CENTRIFUGAL PUMP TROUBLES

A. Symptoms

Symptoms	Possible cause of trouble
Pump does not deliver liquid	1, 2, 3, 4, 6, 11, 14, 16, 17, 22, 23
Insufficient capacity delivery	2, 3, 4, 5, 6, 7, 8, 9, 10, 11, 14, 17, 20, 22, 23, 29, 20, 31
Insufficient pressure developed	5, 14, 16, 17, 20, 22, 29, 30, 31.
Pump loses prime after starting	2, 3, 5, 6, 7, 8, 11, 12, 13
Pump requires excessive power	15, 16, 17, 18, 19, 20, 23, 24, 26, 27, 29, 33, 34, 37
Stuffing box leaks excessively	13, 24, 26, 32, 33, 34, 35, 36, 38, 39, 40
Packing has short life	12, 13, 24, 26, 28, 32, 33, 34, 34, 35, 36, 37, 38, 39, 40
Pump vibrates or is noisy	2, 3, 4, 9,10, 11, 21, 23, 24, 25, 26, 27, 28, 30, 35, 36, 41, 42, 43, 44, 45, 46, 47
Bearing have short life	24, 26, 27, 28, 35, 36, 41, 42, 43, 44, 45, 46, 47
Pump overheats and seizes	1, 4, 21, 22, 24, 27, 28, 35, 36, 41

B. Suction Troubles

1. Pump not primed.
2. Pump or suction pipe not completely filled with liquid.
3. Suction lift too high.
4. Insufficient margin between suction pressure and vapour pressure.
5. Excessive amount of air or gas in liquid.
6. Air pocket in suction line.
7. Air leaks into suction line.

8. Air leaks into pump through stuffing boxes.
9. Foot valve too small.
10. Foot valve partially clogged.
11. Inlet of suction pipe insufficiently submerged.
12. Water – Seal pipe plugged.
13. Seal cage improperly located in stuffing box, preventing sealing fluid from entering space to form the seal.

C. System Troubles

14. Speed too low.
15. Speed too high.
16. Wrong direction of rotation.
17. Total head of system higher than design head of pump.
18. Total head of system lower than design head of pump.
19. Specific gravity of liquid different from design.
20. Viscosity of liquid differs from that for which designed.
21. Operation at very low capacity.
22. Parallel operation of pumps unsuitable for such operation.

D. Mechanical Troubles

23. Foreign matter in impeller.
24. Misalignment.
25. Foundations not rigid.
26. Shaft bent
27. Rotating part rubbing on stationary part
28. Bearings worn
29. Wearing rings worn
30. Impeller damaged
31. Casing gasket defective, permitting internal leakage
32. Shaft or shaft sleeves worn or scored at the packing
33. Packing improperly installed
34. Incorrect type of packing for operating conditions
35. Shaft running off-center, because of worn out bearings or misalignment

36. Rotor out of balance, resulting in vibration.
37. Gland too tight, resulting in no flow of liquid to lubricate packing
38. Failure to provide cooling liquid to water cooled stuffing boxes
39. Excessive clearance at bottom of stuffing box between shaft and casing, causing packing to be forced into pump interior.
40. Dirt or grit in sealing liquid, leading to scoring of shaft or shaft sleeve.
41. Excessive thrust caused by a mechanical failure inside the pump or by the failures of the hydraulic balancing device, if any.
42. Excessive grease or oil in antifriction bearing housing or lack of cooling, causing excessive bearing temperature.
43. Lack of lubrication.
44. Improper installation of antifriction bearings (damage during assembly, incorrect assembly or stacked bearings, use of unmatched bearings as a pair, etc.)
45. Dirt getting into bearings
46. Rusting of bearings due to water getting into housing
47. Excessive cooling of water-cooled bearing, resulting in condensation in the bearing housing of moisture from the atmosphere.

7.15 EXPECTED PUMP TROUBLES AND METHODS TO RECTIFY THEM

	Symptom	Probable fault	Remedy
1	Pump does not deliver liquid	Impeller rotating in wrong direction	Reverse direction of rotation.
		Pump not properly primed – air or vapour lock in suction line.	Stop pump and reprime.
		Inlet of suction pipe insufficiently submerged.	Ensure adequate supply of liquid
		Air leaks in suction line or gland arrangement	Make good any leaks or repack gland
		Pump not up to rated speed	Increase speed

2	Pump does not deliver rated quantity	Air or vapour lock in suction line	Stop pump and reprime
		Inlet of suction pipe insufficiently submerged	Ensure adequate supply of liquid
		Pump not up to rated speed.	Increase speed
		Air leaks in suction line or gland arrangement	Make good any leaks or repack gland
		Foot valve or suction strainer chocked	Clean foot valve or strainer
		Restriction in delivery pipe work or pipe work incorrect	Clear obstruction or rectify error in pipe work
		Head underestimated	Check head losses in delivery pipes, bends and valves reduce losses as required.
		Unobserved leak in delivery	Examine pipe work and repair leak
		Blockage in impeller or casing	Remove half casing and clear obstruction.
		Excessive wear at neck rings or wearing plates.	Dismantle pump and restore clearances to original dimensions.
		Impeller damaged	Dismantle pump and renew impeller,
		Pump gaskets leaking	Renew defective gaskets.
3	Pump does not generate its rated delivery pressure.	Impeller rotating in wrong direction	Reverse direction of rotation
		Pump not up to rated speed	Increase speed
		Impeller neck rings worn excessively.	Dismantle pump and restore clearances to original dimensions
		Impeller damaged or chocked	Dismantle pump and renew impeller or clear blockage
		Pump gaskets leaking	Renew defective gaskets

4	Pump loses liquid after starting	Suction line not fully primed – air or vapour lock in suction line	Stop pump and reprime
		Inlet of suction pipe insufficiently submerged	Ensure adequate supply of liquid at suction pipe inlet
		Air leaks in suction line or gland arrangement	Make good any leaks or renew gland packing
		Liquid seal to gland arrangement logging ring (if fitted) chocked	Clean out liquid seal supply
		Logging ring not properly located	Unpack gland and relocate logging ring under supply orifice
5.	Pump overloads driving unit	Pump gaskets leaking	Renew defective gaskets
		Serious leak in delivery line, pump delivering more than its rated quantity	Repair leaks
		Speed too high	Reduce speed
		Impeller neck rings worn excessively	Dismantle pump and restore clearances to original dimensions
		Gland packing too tight	Stop pump, close delivery valve to relieve internal pressure on packing, slacken back the gland nuts and retighten to finger tightness
		Impeller damaged	Dismantle pump and renew impeller
		Mechanical tightness at pump internal components	Dismantle pump, check internal clearances and adjust as necessary
		Pipework exerting strain on pump	Disconnect pipework and realign to pump
6	Excessive vibration	Air or vapour lock in suction	Stop pump and reprime
		Inlet of suction pipe insufficiently submerged	Ensure adequate supply of liquid at suction pipe inlet

		Pump and driving unit incorrectly aligned	Disconnect coupling and realign pump and driving unit
		Worn or loose bearings	Dismantle and renew bearings
		Impeller choked or damaged	Dismantle pump and clear or renew impeller
		Rotating element shaft bent	Dismantle pump and straighten or renew shaft
		Foundation not rigid	Remove pump, strengthen the foundation and reinstall pump
		Coupling damaged	Renew coupling
		Pipework exerting strain on pump	Disconnect pipework and realign the pump
7.	Bearing overheating	Pump and driving unit strain on pump	Disconnect pipework and realign to pump and driving unit
		Oil level too low or too high	Replenish with correct grade of oil or drain down to correct level.
		Wrong grade of oil	Drain out bearing flush through bearings; refill with correct grade of oil
		Dirt in bearings	Dismantle, clean out and flush through bearings; refill with correct grade of oil
		Moisture in oil	Drain out bearing, flush through and refill with correct grade of oil. Determine cause of contamination and rectify
		Bearings too tight	Ensure that bearings are correctly bedded to their journals with the correct amount of oil clearance, renew bearings if necessary

		Too much grease in bearing	Clean out old grease and repack with correct grade and amount of grease
		Pipe work exerting strain on pump	Disconnect pipework and realign to pump
8.	Bearings wear	Pump and driving unit out of alignment	Disconnect coupling and realign to pump
		Rotating element shaft bent	Dismantle pump, straighten or renew shaft. Renew bearings if necessary.
		Dirt in bearings	Ensure that only clean oil is used to lubricate bearings. Renew bearings if necessary. Refill with clean oil.
		Lack of lubrication	Ensure that only clean oil is used to lubricate clean oil.
		Bearing badly installed	Ensure that bearings are correctly bedded to their journals with the correct amount of oil clearance. Renew bearings if necessary
		Pipework exerting strain on pump	Ensure that pipework is correctly aligned to pump. Renew bearings if necessary.
		Excessive vibration	Refer to Symptom 6.
9.	Irregular delivery	Air or vapour lock in suction line	Stop pump and reprime
		Fault in driving unit	Examine driving unit and make good any defects
		Air leaks in suction line or gland arrangement	Make good any leaks or repack gland
		Inlet of suction pipe insufficiently immersed	Ensure adequate supply of liquid at suction pipe inlet
10.	Excessive noise level	Air or vapour lock in suction line	Stop pump and reprime

		Inlet of suction pipe. Insufficiently submerged	Ensure adequate supply of liquid at suction pipe inlet
		Air leaks in suction line or gland arrangement	Make good any leaks or repack gland
		Pump and driving unit out of alignment	Disconnect coupling and realign pump and driving unit
		Worn or loose bearings	Dismantle and renew bearings
		Rotating element shaft bent	Dismantle pump, straighten or renew shaft
		Foundation not rigid	Remove pump and driving unit, strengthen foundation

7.16 MOTOR FAULTS AND REMEDIES

The following chart is restricted to general faults which should be investigated and rectified by a pump service engineer, with assistance from licensed electrician, where faults of an electrical nature are suspected, which would require the machine to be isolated at the terminal box, a licensed electrician must be called in.

1	Bearing overheating	Pump and driving unit out of alignment	Disconnect coupling and realign pump and driving unit
		Oil level too low or too high	Replenish with correct grade of oil or drain down to correct level
		Wrong grade of oil	Drain out bearing, flush through bearings; refill with correct grade of oil
		Dirt in bearings	Dismantle, clean out and flush through bearings. Refill with correct grade of oil
		Moisture in oil	Drain out bearing, flush through and refill with correct grade of oil. Determine cause of contamination and rectify

		Bearings too tight	Ensure that bearings are correctly bedded to their journals with the correct amount of oil clearance. Renew bearings if necessary.
		Too much grease in bearing	Clean out old grease and repack with correct grade and amount of grease.
2	Bearings wear rapidly	Pump and driving unit out of alignment	Disconnect coupling and realign pump and driving unit. Renew bearings if necessary.
		Rotating element shaft bent	Dismantle motor, straighten or renew shaft. Renew bearings if necessary.
		Dirt in bearings	Ensure that only clean oil is used to replenish bearings. Renew bearings if necessary.
		Bearing badly installed	Ensure that bearings are correctly bedded to their journals with the correct amount of oil clearance. Renew bearings if necessary.
3.	Excessive vibration	Pump and driving unit incorrectly aligned	Disconnect coupling and realign pump and driving unit.
		Worn or loose bearings	Dismantle and renew bearings
		Rotating element shaft bent	Dismantle motor and straighten or renew shaft.
		Foundation not rigid	Remove motor, strengthen the foundation and reinstall motor.
		Coupling damaged	Renew coupling
4	Motor overheating	Motor incorrectly sized for duty	Check design requirements and replace motor if necessary.
		Cooling air ducts/fan chocked	Remove fan cover and clean fan and ducting. On standard protected machines, it may be necessary to strip the motor and clean.
		Amps drawn too high due to under-voltage in supply	Check and rectify incoming supply
		General ventilation in plant room inadequate	Investigate and improve ventilation

5	Ammeter fluctuating during pump operation	Incoming electrical supply inconsistent	Contact supply authority to rectify supply
		Loose connection in system	Check all connections in control panel, starters, terminal boxes
		Stator windings insulation breaking down	Remove cable from motor terminal box and mega test windings and cable separately. Replace cable or have motor windings rectified by rewinding, reinsulation, backing.
		On slipring motor brushes not contacting sliprings correctly	Expose sliprings and check and rectify brushes, brush holders, pressure springs.
		Ammeter faulty	Check and replace if necessary.
6.	Motor will not start	Supply voltage not reaching motor	Investigate and rectify
		Connections in system faulty or cable broken	Tighten and clean all connections and check cable for continuity
		Motor incorrectly connected	Check connections against diagram
		Motor incorrectly sized for application	Try starting motor uncoupled. Check sizing of motor to requirements of pump. If necessary change to motor more suited to requirements *i.e.*, high starting torque, etc.
		Starter contactor faulty	Check and rectify
		Auto transformer faulty	Check and rectify
		Brushes not contacting sliprings correctly	Expose sliprings and check and rectify brushes, brush holders and pressure springs.
		Internal connections in motor faulty	Check and rectify
		Rotor resistance incorrect	Check design and rectify rotor
		Overloads in contactor incorrectly adjusted	Check and reset to required value

7	Motor starts but will not take load	Overloads in contactor incorrectly adjusted	Check and reset to required value
		Supply voltage incorrect	Check and rectify
		Lose connections in system	Check and rectify
		Motor incorrectly sized for duty	Check design and rectify
		Method of starting incorrect for duty requirement	Check design and rectify
		Motor design incorrect	Check design and rectify
		Driven unit excessive stiff	Check and rectify
		Direction of rotation incorrect	Check and rectify
8	Brushes on slipring motor spark continuously	Motor being overloaded	Check design and rectify
		Brushes installed are the wrong grade	Check with original supplier/ manufacturer
		Brushes sticking in holders	Clean and refit brushes and check pressure springs
		Brushes need bedding	See supplier/manufacturers handbook
		Sliprings dirty/worn	Check and clean, restore to original design condition replace brushes
		Machine vibrating excessively due to poor bearings, alignment, foundations	Replace bearings, check alignment, repair/rebuilt foundation
9	Motor stator windings burnt out	Moisture enters windings	Fin probable point of entry and plug, have stator rewound
		Condensation forming due to anticondensation heaters not working	Check a/c heaters for operation and sizing

		Overloads set too high and/or motor too small for job	Check overload settings and design for sizing motor
		Design of motor incorrect for application, *i.e.*, SPDP instead of TEFC, thus allowing moisture entrance	Check suitability of design to actual installed item
		Class of insulation not suited to application, *i.e.*, class E instead of B or F etc.	Check suitability of design to actual installed item

7.17 PREVENTIVE MAINTENANCE TECHNIQUES

In most major industries, failure of an essential item of equipment can result in a total emergency shut-down of the complete plant. It is for this reason that on the more vulnerable possible points of failure, such as rotating machinery, it has become normal practice to employ a working machine with a 100% standby machine standing alongside to take over the function in the event of failure occurring. If such an occurrence takes place, immediate action must be taken to restore the damaged unit to service with minimum down time to ensure continuity of service in the event of a repeat occurrence.

Regular Routine Checks

Despite the above precautions, regular checks on all the more vital items of equipment, including almost without exception any installed pump units should be made to exception any installed pump units should be made to identify in good time, possible developing problems which will have to be faced at the first opportunity.

7.17.1 AIDS TO PREVENTIVE MAINTENANCE

Fixed Instrumentation

Any fixed instruments may in themselves create an additional possibility of failure and require very careful observation to ensure such instruments are in fact functioning correctly: Pressure gauges—suction delivery and flow meter.

7.18 PUMP DESIGN

A large variety of pumps are available; the most common are the vane, gear, centrifugal, piston and plunger type pumps. Gear, vane and piston type pumps are often used in oil hydraulic systems and as oil is the medium being pumped, this provides the necessary lubrication. Where water or chemicals are pumped, lubrication is provided through alternative methods. Pumps are manufactured using different types of bearings which are one of the most critical components of pumps. For instance, a worn pump may still supply a certain flow but if the bearings are worn or damaged, the pump will probably seize. The correct type of oil or grease is therefore crucial in maintaining the efficiency of the pump.

7.19 PUMP LUBRICANTS

7.19.1 FUNDAMENTALS

As with all other applicants, the same set of fundamentals apply to pump lubrication. The first fundamental involves the choice between an oil and a grease. Once this has been established, other aspects must be considered. These aspects include:

- Correct viscosity grade or grease consistency.
- Compatibility of a lubricant with components, such as pump packings.
- External contributing factors, such as temperature extremes.
- Presence of water or chemicals.
- Length of service of the lubricant.
- Seal compatibility.

All these factors play a significant role in the efficient working function and life of a pump.

7.19.2 FUNCTIONS OF A LUBRICANT

Lubricants must provide sufficient film to separate moving parts, but they also have other vital functions. Firstly, they must protect the pumps internal components from rust and corrosive pitting. Not only does corrosion produce pitting. But the oxide itself is abrasive. The lubricant must therefore be chemically protective and non-corrosive. Secondly, the lubricant must act as a heat transfer to maintain an even temperature throughout the bearings. Furthermore, the lubricant must protect the pump and bearings from contamination such as water and dirt, and is often considered to be the second line of defense of the bearings.

7.19.3 CHOICE OF LUBRICANT

The choice of lubricant invariably lies between grease and oil. The effectiveness of grease, which is widely used for pump lubrication, is based on the amount of oil released by the grease for service.

The decision of lubricant is also very dependent on:

- bearing type
- bearing size
- rotational speed of bearing
- loading
- operating temperature
- exposure to chemicals
- sealing requirements.

In most cases, the republication of the bearings and other components of the pump must be considered, especially where extreme factors play an influential role.

7.20 BEARING LUBRICATION

In theory bearing elements are considered to have line or point contact. However, in practice, this condition is virtually impossible to attain as the unit pressure would be infinitely high. Even the hardest steel is deformed under pressure. Each of the two loaded surfaces will deform until the contact area between them is sufficient to support the applied load. The extent of deformation is dependent on material strength, applied load and surface conformity. For plain bearings, normally manufactured of white metal and brass, the lubrication principle follows that of hydrodynamic law. This is achieved where the rotation of the shaft in the bearing causes an oil wedge to form with the divergence in common with the direction of rotation of the shaft. This allows the shaft to actually "climb" onto the oil wedge, thus separating the bearing and the shaft. This rule also applies generally to rolling type bearing elements.

7.21 GREASE LUBRICATION

Although lubricants in the form of oil are undoubtedly the most efficient, grease has many advantages due to its constituents. Most greases consist of a thickener, normally a metallic soap or a combination of soaps and lubricating oil. The oil used in a grease depends on the required viscosity at a given working temperature and the operating conditions. Grease is retained easily within its application and is therefore particularly advantageous in terms of cleanliness, especially in the food and textile industries. By virtue of its semi-solid state, it tends not to drain away from a point of lubrication and also assists the seals in performing more efficiently. Although relubrication is relatively quick and easy, caution must be taken to avoid over or under lubrication. If bearings are over-packed, the grease can obstruct raceways and bearing elements, causing frictional resistance and overheating. This causes breakdown on the fibrous structure of the grease, resulting in loss of oil and hence poor lubrication. The thickening agents are used to make a grease vary in fiber length. These fibers form a three dimensional structure which absorbs or holds the lubricating oil, much as a sponge will absorb retain water. In use, oil is gradually released to perform its lubricating function. Most greases are made with lithium, or calcium soaps as thickeners. Calcium greases are relatively low in cost and used where operating speeds and temperatures are not too high. Calcium greases are resistant to water, can be enhanced by the addition of anti-corrosion additives and are suitable for operating temperatures between 50°C and 90°C. Sodium based greases are not very water tolerant, although they can absorb a certain percentage of water before the lubricating efficiency is affected. This type of grease can be used up to temperatures of approximately 90°C. Lithium and lithium complex greases are becoming increasingly common in pump lubrication as they are resistant to water, to the extent of not being washed out even by large quantities of water. Lithium based greases can also tolerate higher working temperatures than calcium and sodium based greases and are considerably more stable in high speed, high load conditions. Generally lithium based greases can function satisfactorily at temperatures in excess of 120° C. Aluminum complex greases are also becoming more common, especially where the presence of water is a factor. These greases can withstand a high degree of water contamination and can cope well with high temperatures. The operating speeds of the bearings/shafts are also tolerated while using aluminum type greases. Most greases contain certain additives to enhance their capabilities, the most common being anti-oxidant, anti-wear, extreme pressure and anti-corrosion additives. Some greases contain a tacky additive to reduce dripping or fling-off during operation. Greases containing tacky additives can cause channeling at high speed which can, in turn, cause premature bearing wear. Many greases also contain "dry additives" such as graphite and molybdenum disulphide, which acts as a substitute lubricant if the grease itself is forced out of its lubricating area. Caution must be exercised when using lubricants with graphite and molybdenum disulphide in rolling element bearings, as a high concentration of these additives can cause rolling elements to slide instead of roll thereby producing a flat surface on a ball or roller. This type of additive is usually used in greases intended for shock loaded or high impact applications such as found in earthmoving machinery. Grease replenishment is subject to operating conditions. Certain bearings are pre-packed in the manufacturing process and "sealed-for-life". As a general rule the faster the rotational speed and the higher the operating temperature, the more frequently the grease should be replenished. The frequency of replenishment is normally stated by the pump manufacturers.

7.22 OIL LUBRICATION

Oil lubricants, in most cases, is the most convenient and effective method of lubrication. Where ambient temperatures and speeds are too high for satisfactory grease lubrication, oil or an oil mist must be used. Where speeds are not too high, the simplest method of oil lubrication is an oil bath. To avoid excessive churning of the oil, the oil level in the bath should not be higher than the centre of the rolling elements at their lowest point. When an oil bath is used with tape rolling bearings, a centrifugal effect pumps oil from the small to the large ends of the rollers. This effect allows a constant flow of cool oil through the bearings, thus eliminating the need for an external pump feed. At high speeds, it is necessary to reduce the amount of oil in the bearing housing while, at the same time, maintaining sufficient lubrication. In this case, wick or drip feeds are commonly used. For heavy loadings and very high bearing speeds, oil force fed from a sump or reservoir is the most suitable. The choice of oil is critical in pump lubrication. Some oils are designed to absorb a certain quantity of water while others are designed to reject water completely. Leakage of certain chemicals into the lubricant during operation, can react with the additives incorporated in the oils formulation, causing potential problems. The most important factor when choosing a lubricant is the oil viscosity. Logically, an oil which is too heavy or too light will cause damage, not only to the bearings but also to the pump components. Pumps such as vane pumps, internal and standards gear type pumps, piston pumps and certain plunger type pumps rely on the pumped medium to lubricate the moving parts. A common rule to follow is that the faster the components move, the thinner the oil should be. The viscosity and the additive requirements are stated by the individual pump manufacturers and have been well researched and tested before recommendations were made. Most oil lubricated pumps where the medium being pumped is not the pump lubricant, are normally lubricated with a hydraulic or turbine type oil. These oils contain anti-corrosion additives and, in most cases, do not require the use of the extreme pressure additives, as would be found in gear oils. Engine oils commonly used in pump lubrication as they absorb a fair amount of water, can tolerate heat, and contain the necessary additives in the form of anti-corrosion and anti-wear additives. In conclusion, maximum pump life and efficiency are greatly dependent on correct lubrication and cleanliness. If in doubt, do not take a chance–rather ask.

7.23 PRIMING

A centrifugal pump located above its suction level is kept full of liquid and ready to pump on start-up. Instead of depending on a foot valve, a vacuum is drawn in the pump casing. The priming system is connected to each centrifugal pump through a vacuum priming valve. This is a float valve which closes when liquid rises in it. Several connections may be needed to evaluate air from all high points in the centrifugal pump's suction eye and volute. The priming valve is tapped at three places for these connections.

7.23.1 Dry Priming Life

As the liquid level in the suction pipe rises, the differential pressure across the pump increases until, at a given pump speed n, air slip ΔQ equals pump displacement Q_n. There is therefore no further increase of the vacuum (negative gauge pressure) above the liquid level in the suction pipe. The atmospheric pressure is unable to raise the liquid level any further and the maximum suction height (priming life h_1) at pump speed has been reached. It has been shown that volumetric efficiency increases with pump speed and it follows that this will also increase the effective suction height (h_2). However, as suction height increase still further, the differential pressure will lift the resilient impeller blades from the impeller bore, resulting in additional slip around the periphery.

7.24 MAINTENANCE PRACTICE AND FAULT FINDING

7.24.1 GENERAL

The causes of malfunction of the centrifugal pump are either mechanical inadequacy or the incorrect selection of the pump type and/or pump size or any of its components. In the latter event it is possible that the pump selection has been incorrect from the beginning or the plant operations have been changed or have deteriorated. Mechanical pump problems, which may result in mechanical failures or reduced hydraulic performance, can in most cases be rectified easily by replacing the respective components. Hydraulic and performance problems are often caused by incorrectly assessed plant conditions or the deterioration of one or more plant components (possibly mechanical components can be prevented and/or postponed by correct maintenance). Under the heading Erosion a wide range of pump problems with the possible causes and remedies has been summarised. Similar lists have been compiled by most pump manufactures, as well as a number of textbook authors. These lists of possibilities are a help for the novice in this field; however the pump engineer adopts a more basic approach by comparing the actual with the intended operation conditions. With the knowledge of the internal function of the pump, he investigates the possibilities which can cause deviation from the design conditions. This investigation can be done by tests and/or calculations. If this pump fails to operate, either the failure is of a mechanical nature, or the expected hydraulic performance cannot be obtained. Both failure types could be caused by a mechanical deficiency or a hydraulic problem. Locate the cause of the failure, in order to avoid re-occurrence; the basic approach for these two main sources of failure is covered as follows.

7.24.2 MECHANICAL FAULTS

Mechanical faults and their causes are best assessed and evaluated by physical inspection. The pump should be dismantled carefully and all damage as well as the position of broken components should be noted. Mechanical damage can be divided into the following groups all which can influence on the mechanical running behaviour and/or the hydraulic performance of the pump.

Breakages

From the fractured sections it is almost always possible to determine if the breakage has occurred due to material flaws, insufficient or unsuitable materials, straight-forward replacement with the correct components will solve the problem, while in cases of fatigue failures, hydraulic conditions may have to be taken into account.

Score marks

The position of these marks can give indications to incorrect assembly or manufacture (eccentricity, incorrect tolerances and clearances unparallel fits and seats etc.) or the behaviour of the pump rotor. The axial and radial hydraulic forces on the pump rotor are affected by the unfavourable and unplanned operating conditions. The direction of the score marks will give an indication as to which forces have caused them.

Erosion: Abrasion

The material loss due to abrasion depends on the velocity size, hardness and angle of impact concerning solids in the pumped liquid. Abrasion does not only reduce the wall thicknesses of casings and impellers, but also increases clearances and throttle gaps. The latter will result in unstable running behaviour of the pump rotor, as well as an increase in the axial hydraulic thrust and an efficiency loss.

Cavitation

From the position of the cavitation damage it can be determined, whether cavitation occurred during part load or overload operation.

Corrosion

While abrasion damage always shows smooth surfaces, corrosion damage results in rough or even pitted material surfaces. Almost all surfaces are affected. In high velocity areas, material will be damaged by corrosion –erosion effect. In these areas, high flow velocities will destroy the passivation layer which most materials form after a certain period of operation. This passivation layer slows down the corrosion processes when left undamaged. Clearances between high and low pressure areas are always subject to high flow velocities and corrosion has a much greater effect. As these clearances often act as hydrodynamic bearings for the pump rotor, the running behaviour of the rotor is also influenced.

7.24.3 Hydraulic Faults

A pumping plant is expected to move a certain volume of water from one position to another. If this volume is not achieved the plant conditions have not been selected or the pump does not perform as expected, in order to establish the cause of the non-performance. It is best to compare the present operating conditions with the intended ones. If a lower flow rate is experienced, the head generated by the pump should be measured with closed, fully open and partly opened discharge valve, and compared. The simple site tests do not provide conclusive results. A pump should be tested in the more ideal conditions of the test plant.

Valve closed: as per pump curve

Valve open: as per pump curve, but higher than the calculated duty point.

Valve partly open: as per pump curve

The system conditions are different than originally calculated (*i.e.*, blockage, pipe friction, filters, strainers etc.)

Valve closed: as per pump curve

Valve open: lower pressure of the system curve

Valve partly open: Pressure lower than pump performance curve.

This condition indicates a pump fault (blockage in the impeller, diffuser or casing etc.)

Valve closed: as per pump curve

Valve open: pressure lower in the vicinity of the system curve

Valve partly open: as per pump curve

This indicates a performance decreased due to insufficient suction pressure (cavitation).

Valve closed: lower than pump curve

Valve open: lower than pump curve in the vicinity of the system curve.

Valve partly open: lower than pump curve

This condition indicates an incorrect speed or impeller diameter. The reason could also be the incorrect direction of rotation. A pump impeller generates very low head of liquid column when operated in the wrong direction.

5.3 REGULATION BY DIAMETER CONTROL (OUTER DIAMETER TRIMMING TO MEET THE DESIRED [illegible])

Index

t stands for table.